F-111

U.S. AIR FORCE

F-111

Bill Gunston

Charles Scribner's Sons
NEW YORK

Printed in Great Britain
Library of Congress Catalog Card Number 77-95199
ISBN 0-684-15753-5

Contents

Preface

Everybody interested in aviation knows that the F-111, formerly called the TFX, is the greatest combat aircraft there has ever been (or the worst). Everyone on Capitol Hill in Washington knows that the story of its contract award and development for operational service was the biggest scandal of modern times (or the biggest success story). Those who created it ought to feel inflated with pride (or ashamed). No weapon system in history has done so much to destroy (or build) personal reputations. No weapon system has ever proved so immensely valuable (or been so great a disappointment). It can be summed up quite easily:

'The TFX program has been a failure'. (*Report on the TFX Program* by the Committee on Government Operations of the United States Senate).

'The TFX program has resulted in the most unstoppable tactical airplane in the world'. (Gen John D. Ryan, USAF Chief of Staff).

'The F-111 is a lemon'. (general consensus of popular media — press, radio and TV — throughout the United States, 1965-75).

'The F-111A is the best all-weather interdiction aircraft in the world' (Lt-Gen J. W. Carpenter III, USAF Assistant Vice-Chief of Staff).

'TFX stands for Terrible ******* Experience, in my book'. (Ben Schemmer, Editor, *Armed Forces Journal,* 1965).

'The F-111 is our most sophisticated tactical aircraft, and is easily the world's most effective aircraft for the long-range tactical attack mission'. (Richard C. Seamans, Secretary of the Air Force).

'All it took to turn a critic and a coward into a convert was a few hours at Nellis flying that airplane with the men who trust their lives to it every day'. (Ben Schemmer, Editor, *Armed Forces Journal,* 1972).

So, obviously, the reader can see that, whereas everyone knows that the F-111 is a disaster, the people who actually live with it, fly it, and have been to war in it, think exactly the opposite. I am glad to have been asked to put together this book. I was in the fortunate position of being a neutral on the sidelines. The critics of the programme certainly had a case, and it was sheer bad luck that — quite independently of the original controversy over 'commonality' between Air Force and Navy versions, and who should build it — the F-111 design suffered more than any other aircraft of recent years from such deep technical faults as excessive drag, excessive weight, gross malfunction of the propulsion system, poor flight performance and manoeuvrability, catastrophic structural failures and escalation in costs. Beyond doubt, there was mismanagement of the project at the highest levels in the US government, and too much determination to bulldoze ahead with aircraft that were unsatisfactory or deficient in varying degrees for their missions. This sadly tended to polarise opinion into hardened 'antis' and hardened 'pros'.

It took years for the passions to cool, and for objectivity to be restored. Maybe today is the earliest time at which one could write a fair and objective book on the Aardvark — which is what the chaps call the One-Eleven — and I hope this book will be considered to be just this. Those who know the worth of this truly great aircraft may not like the early chapters; they may think I am merely raking over old arguments, re-opening old wounds and flogging dead horses. But they are all part of the story; they cannot be left out. The once-vitriolic 'experts' who waxed rich by knocking the One-Eleven are today much less in evidence; like Concorde, the F-111 is not quite the front-page-stopper it once was. If there are any still in business, they will not like the rest of this book, which makes it abundantly clear that the F-111 in most versions is neither a failure nor a lemon.

It is not generally appreciated that the very first version to go into service, the F-111A, is the only really modern military aircraft to have gone to war. It had two bites at this cherry, and on its second trip to south-east Asia in 1972-73 flew more than 4,000 combat missions under incredibly tough conditions and in the face of the most intense anti-aircraft defences ever assembled. Those who are influenced by the statements of politicians, professional critics and others who have never actually touched a One-Eleven might do well to give equal weight to the statements of the men whose job it is to fly the One-Eleven professionally. They called their automatically guided terrain-following flight 'skiing'. Skiing is an exciting sport, but how many have tried it at Mach 0.9 in foul weather with the invisible mountains whipping past overhead? The Aardvark jocks were the first in the world to go to war in this adrenalin-pumping way. They were unstoppable.

Soon there will be answers to the lone sharp-nosed skier. By the next century the Aardvark will no longer be in the front line. But it will be one of the famous aircraft of history, not so much because it was the first with swing-wings, or the first with afterburning turbofan engines, or the first tactical aircraft to weigh 50 tons, but because it was the first aircraft in history that could go out in thick fog at night and drop a bomb in a 50-yard circle, ten times out of ten. That is still quite something, especially as the circle can be a thousand miles away. Lemon? Failure? Then show me a success.

Bill Gunston
Haslemere, Surrey

Acknowledgements

The background to this book was picked up from many people in 1959-64 in the formative years of the programme. In 1977 in writing this book I would especially like to thank the following for their help: Jay Miller, Editor *Aerophile* of Austin, Texas, who went beyond the line of duty in providing valuable material and prodding others; Frank Strnad, professional photographer; Grover 'Ted' Tait, of BDM Corporation at Edwards AFB; Rob Mack of GD Fort Worth; Harry Rayner of DoD Canberra, Australia; Henri H. Simons, v-p (Europe) of Grumman International; and many past and present members of the 20th TFW, US Air Force.

The final chapter is from *Flying Combat Aircraft of the USAAF-USAF* (Iowa State University Press), and the author is Maj Peter M. Dunn who flew more than 100 missions over North Vietnam.

Glossary

The following are some abbreviations commonly used in this book:

AB	airbase (USAF base in foreign territory).
AFB	Air Force base.
BW (M)	Bomb Wing (Medium).
CAP	Combat Air Patrol.
DoD	Department of Defense.
ECM	Electronic countermeasures.
FAD	Fleet Air-Defense.
FLIR	Forward-looking infra-red.
GD	General Dynamics (US corporation).
Hi	High altitude, over 30,000 feet.
IR	Infra-red (heat).
Lo	Low-altitude, down to 200 feet.
Pacaf	Pacific Air Forces.
RAF	Royal Air Force.
R&D	Research and development.
RDT&E	Research, development, test and evaluation.
RFP	Request for Proposals
SAC	Strategic Air Command.
SAM	Surface-to-air missile.
SEA	South-east Asia.
SecDef	Secretary of Defense.
TAC	Tactical Air Command.
TFR	Terrain-following radar.
TFS	Tactical Fighter Squadron.
TFW	Tactical Fighter Wing.
Triple-A	Anti-aircraft artillery.
WSO	Weapon-system operator.

Below: An F-111E with clean wings at 26° in a gentle turn. / *USAF*

The Glittering Prospect

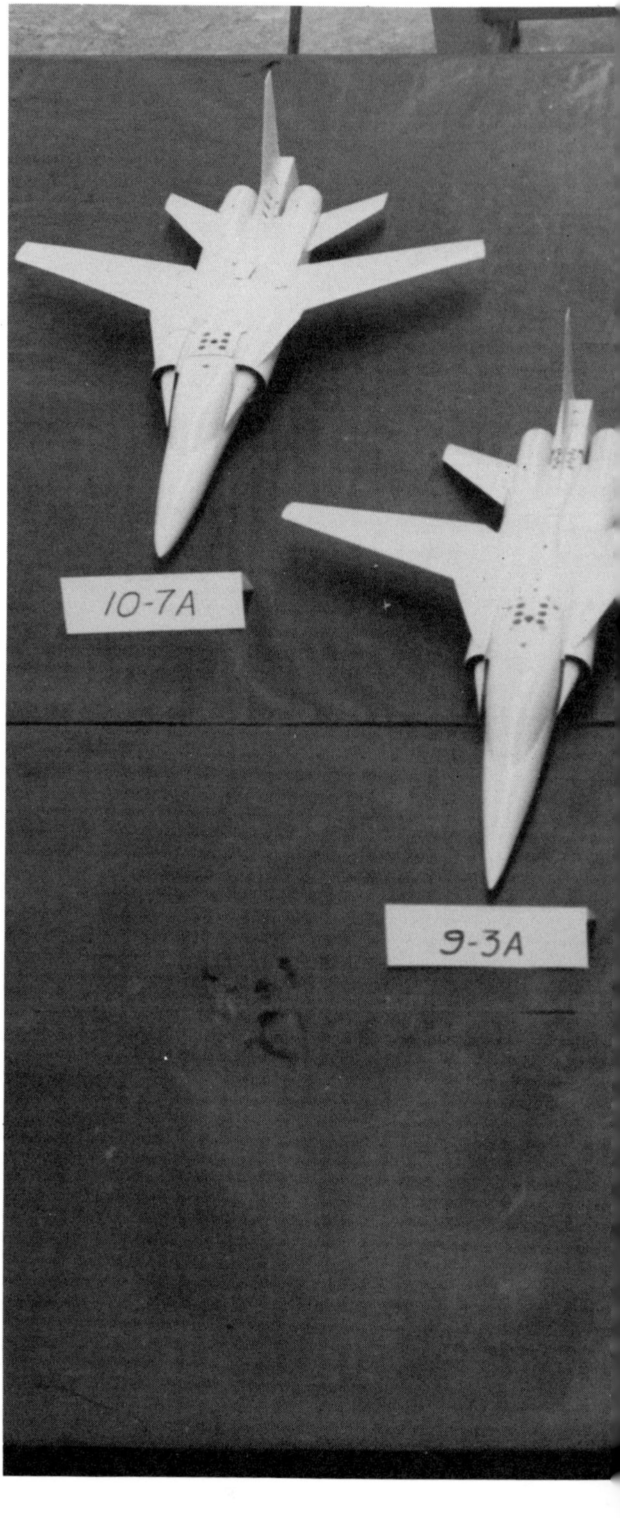

On 20 January 1961 a new man was appointed to one of the biggest jobs in the world. President Kennedy had asked Robert S. McNamara to be his Secretary of Defense. What are the qualities that fit a man for such a titanic post? McNamara had an impressive reputation as a large-scale manager. He had taught management at the Harvard Business School. He then became a vice-president of the Ford Motor Company and, in unscrambling its chaotic administration, put that enterprise back on its feet. Kennedy felt that the Pentagon could do with such a man.

So did about 19 out of every 20 Americans. The Pentagon, like most large organisations that cannot readily answer back, had become synonymous with creaking bureaucracy, hidebound inefficiency and armies of people whose salaries were more obvious than their output. McNamara, like other Americans, had inevitably concluded that the whole Department of Defense (DoD) was due for an almighty shake-up. Before long he had begun to shake, setting a personal example of a 12- to 15-hour day, often seven days a week, while he constructed his own environment of what became called 'Whiz Kids'. Most of these were younger editions of himself, brilliant young experts on business management, computers, finance, engineering, the law, mathematics and the other software tools the new SecDef needed. Before long they had become heartily disliked by silver-haired men of great eminence — captains of industry or generals with four stars — whose knuckles had been sharply rapped in falling foul of the new streamlined defence machine.

Right: Tunnel models of November 1961 GD TFX configurations, all at minimum sweep except Shape 500 which is at maximum sweep of almost 90°. Dark spots on the models are pressure sensing points. / *GD via Jay Miller*

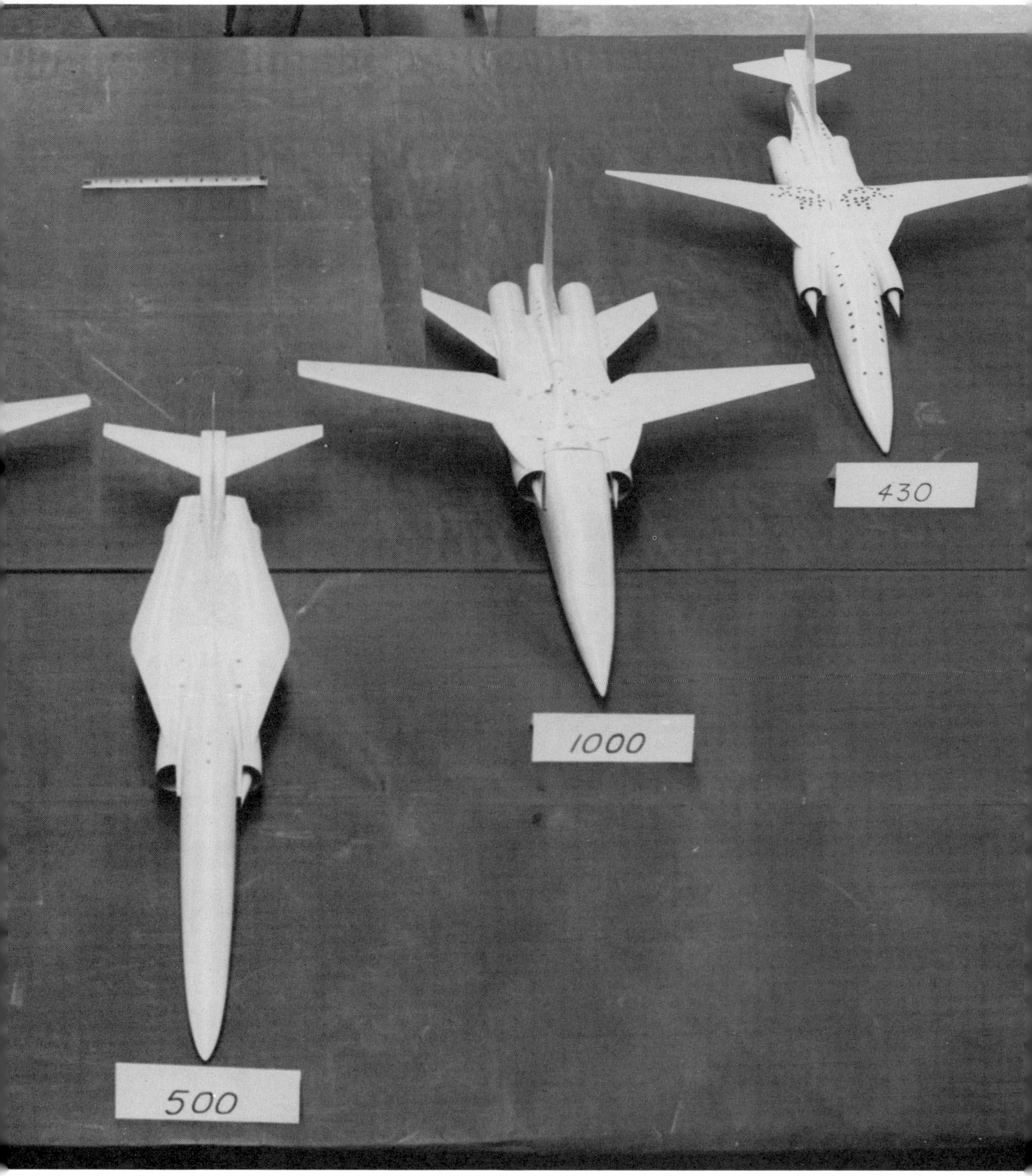
430
1000
500

In between whiles McNamara fitted in detailed briefings on the current procurement programmes. Among the many major weapon systems then being procured or studied were two that seemed to have much in common. One was to meet Air Force Specific Operational Requirement 183 (SOR-183); the other was for a Fleet Air Defense Fighter for the Navy. Both had a background history of several years.

SOR-183 had been an outgrowth, on 14 June 1960, from studies begun in 1958 at the NASA Langley Laboratory, which in turn had been influenced by earlier work by Vickers-Armstrongs (Aircraft) at Weybridge in England. These studies had centred on the variable-sweep 'swing wing'. Even as early as World War 2 German aerodynamicists had realised that for ideal transonic performance an aeroplane needs to have wings mounted on pivots. For long range at low speed, or for short take-off and landing, it needs to have long slender wings stretched out to a wide span on each side. For supersonic performance it needs wings that are swept back on each side, while for comfort at supersonic speed in a low-altitude attack, in dense air, it needs wings swept back at the sharpest possible angle, giving an extremely small span. But making a good swing wing proved elusive.

Throughout the 1950s combat aircraft had either swept wings, or 'delta' wings with the rear part filled in to form a triangle. All these were structurally reasonable, gave a low ratio of thickness to chord (distance from leading edge to trailing edge) and moderate transonic and supersonic drag; but they were inefficient during take-off and landing, or in subsonic long-range cruising flight, they made it difficult to fit powerful high-lift flaps, they forced the aircraft into an extremely nose-high attitude for maximum lift (so that the pilot had a poor forward view when landing, or when formating on a slow tanker aircraft) and they introduced unpleasant problems with rolling and sideslipping in the high-lift regime. Designers had no choice but to accept these faults, and others, if the speeds that could be attained with jet engines were to be realised. Pilots had no choice either, and tactical air forces whose airfields might be blasted in the opening minutes of any future war found themselves equipped with aircraft that needed ten thousand feet of concrete.

Of course, everyone recognised the gravity of the problem, but technology did not exist to find solutions. One possible answer seemed to be the Zell technique (ZEro-Length Launching). This shot tactical aircraft into the sky by sheer brute force. Trials were flown with such important long-runway machines as the F-100 Super Sabre and F-104 Starfighter. Fearsome solid-propellant rockets were clipped underneath, the aircraft mounted on an inclined launcher with its wheels clear of the ground, and the pilot provided with

a long ladder and an extra button in the cockpit (and maybe a prayer-book). Zell worked — impressively — but it had its disadvantages. Zelled air forces would have to spend vast sums on the rocket motors, each of which would tax the capacity of an average fork-lift truck in moving it, resulting in severe logistics problems. Each motor would be used once (though it might be refurbished and refilled), adding greatly to the cost of each mission. And Zell did nothing to help aircraft land again after their mission on a cratered airfield.

Another answer was jet lift, or some other scheme giving V/STOL (Verical or Short Take-Off and Landing). Many millions of dollars were spent investigating schemes that used some or all of the installed engine thrust directly in lifting the aircraft, but none seemed to offer much promise. In the event, of course, one of these schemes — the simplest of all, using a single engine with vectored nozzles — has led to a most effective combat aircraft, but 20 years ago this did not seem likely. In any case, all the V/STOL schemes so compromised the aircraft as to make it difficult to achieve the desired flight performance, especially the range.

Swing wings, on the other hand, exacted only a very small penalty in structural weight and complexity, and hardly bothered the supersonic performance at all — if only they could be made to work. There had been one or two attempts to build swing-wing aircraft prior to SOR-183, notably the Grumman XF10F Jaguar Navy fighter of 1952, but the game had seemed not worth the candle. There were severe problems of mechanical design, and the trim problems caused by the forwards or rearwards shift of the aerodynamic

Above left: Nose jettison schemes for Navy and TAC TFX proposals dated February 1962. Later GD settled for a crew capsule recessed into the upper part of the fuselage.
/ *GD via Jay Miller*

Above: GD/Grumman TFX configuration 10-7 which formed the basis for the First-Round proposal. This 1/25th scale model was dated 1 December 1961.
/ GD via Jay Miller

centres of the pivoting wings were not easy to solve in an elegant way. But in 1954-55 the Vickers research staff had discovered an attractive arrangement in which the left and right wings were pivoted well out from the aircraft centreline. The pivots were on the extremities of a large delta-shaped inboard fixed wing, called a glove, which unfortunately spoilt the airflow over the tailplane and reduced the gains with the wing in the unswept (spread) position, but was needed to make the aircraft work. Without a glove the part of each wing inboard of the hinge would be too long to be housed inside the fuselage in the unswept position. NASA's Langley Laboratory was given the British data and went on to take the idea much further. Under brilliant aerodynamicist John Stack a complete solution was found in early 1959 to all the basic problems.

Stack was no ivory-tower researcher. He was in daily contact with industry, airlines and the military, and soon saw with great clarity that the most immediate application of his laboratory's swing-wing solution could be a new multi-role tactical fighter. He had no doubt that it would be of great benefit to the US Navy with various categories of carrier-based aircraft, but the prime customer seemed to be Tactical Air Command (TAC). Numerically the largest of all USAF commands, TAC was in 1959 equipped with such aircraft as the F-100, F-101 and F-105 — good and tough workhorses that could deliver nuclear or conventional ordnance over useful ranges and which, when clean, could accelerate to highly supersonic speeds. But TAC's Commander, General Frank F. Everest, told Stack that his force was equipped with aircraft whose technology should have been superseded (though in practice it had not been). Many things had been done since the early 1950s when the TAC aircraft had been designed. In August 1959 Stack came to see Everest at the latter's headquarters, conveniently nearby at Langley AFB (like the NASA laboratory, named for a pioneer of heavier-than-air flight). They discussed the prospect, and identified the following advances:

1 First and foremost, the swing wing looked promising at last, and its promise was enormous. For the first time a tactical aircraft could combine heavy-bomber payload, heavy-bomber range, fighter-like speed, outstanding manoeuvrability and the ability to operate from short airstrips.
2 Progressive development of high-lift systems, involving refined wing slats and flaps, could readily be applied to the swing wing and used in the spread position, thereby reducing field length still further. This could not be done with swept and delta wings.
3 Great strides in engine technology, pioneered partly in Britain, had led to the by-pass jet or turbofan in which the air delivered by the low-pressure

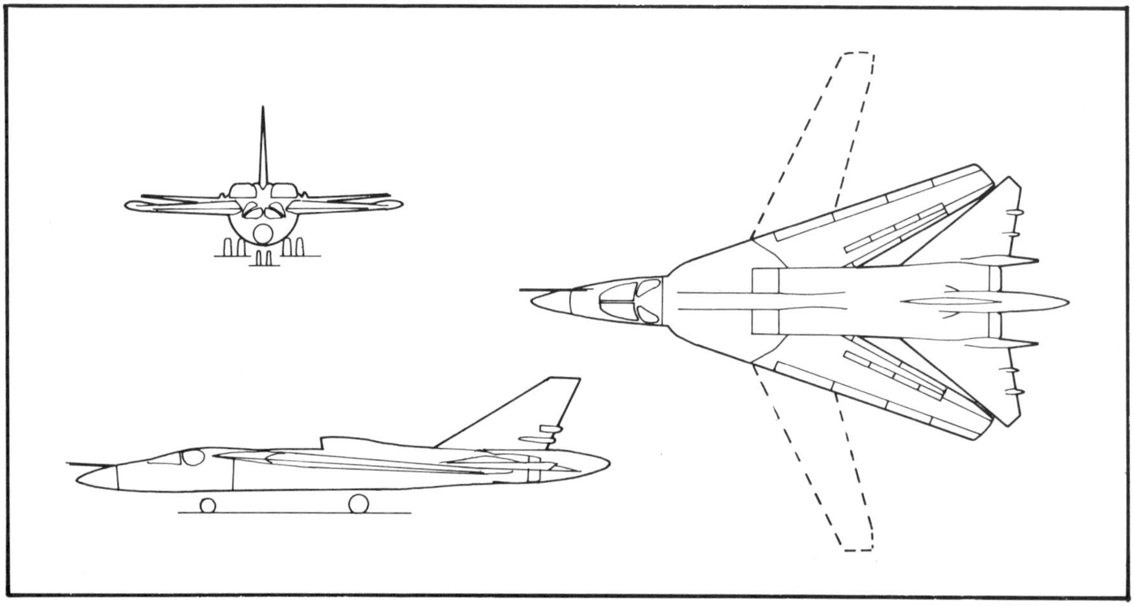

compressor was divided, part going through the 'core engine' (high-pressure compressor, combustion chamber and turbines) and the rest being discharged through a surrounding duct to the propelling nozzle. This promised much better aircraft range, by reducing the fuel consumption, while at the same time allowing the extra thrust obtained by afterburning (injecting additional fuel into the jetpipe) to be greatly increased. No afterburning supersonic turbofan existed in 1959, but it looked a straightforward job for the engine manufacturers.

4 New materials were becoming available for different parts of the airframe, and the useful metal titanium had acquired so much experience in secondary (non-critical) parts of aircraft that it was now ready for use in primary structure.

5 By no means least, the shortcomings of the first generation of supersonic tactical aircraft had led to an illuminating dossier of how things could be done better in the aircraft that would replace them. In particular, it was clearly possible to design the densely packed internal systems, and especially the avionics (aviation electronics), so that they would test themselves automatically, indicate any faulty item, and be so arranged that offending items could be removed in seconds and replaced by ones known to be serviceable. Previously, combat aircraft had been designed like so many private cars are even today, with numerous important but fault-prone components which could not be removed without dismantling half the aircraft. In 1959 the possibility of changing virtually anything with no more trouble than swapping over two plug-in boxes offered a tantalising prospect.

Above: Manufacturer's three-view of the Boeing Model 818, the Wichita submission for the TFX attack version for the USAF. This was the final Boeing proposal, with TF30 engines. Empty weight was given as 35,410lb and basic mission gross weight 69,000lb.
/ *Kurt Miska collection via Jay Miller*

It was with such thoughts in mind that Everest convened a formal meeting of Air Force commanders and senior engineers, which, in turn, triggered off further conferences at Wright-Patterson AFB (the Air Force technical headquarters), the Pentagon, Langley Lab., and selected aircraft manufacturing companies. By 14 June 1960 the result had been printed in the concise form of Air Force specification SOR-183. It was possibly the most challenging ever issued up to that time. Perhaps those who drafted it were carried away by the exceptional list of major technical advances that could be cranked into the new aircraft, because they set their sights very high indeed.

In 1960 the best fighter-bomber in TAC was the Republic F-105 Thunderchief, and Everest's men began by stipulating the F-105 mission as a minimum requirement. This implied supersonic speed at all flight levels, and internal carriage of nuclear weapons. One of the more obvious shortcomings of the well-loved 'Thud' was its long field-length. An analysis of F-105 operations in Europe revealed that there were no overruns or barrier crashes on a 10,000 foot base in West Germany but a distressing succession at an 8,000 foot base a mere six miles away. As a ball-park figure Everest said he wanted the new aircraft to need 'half the distance of the 105', and when SOR-183 was finally drafted this had been refined to 3,000 feet. On the matter of flight performance the document adhered to Everest's '1,700mph' by specifying a Hi (high-altitude) dash speed of Mach 2.5. Everest was especially demanding in the matter of range. He wanted a treetop (Lo, or 200-250 feet) dash at 1,000mph for 400 miles, and this eventually appeared as Mach 1.2 with internal weapons over the final 200 nautical miles (230 miles) to a ground target at a distance of 800 nautical miles, with an objective of flying the whole outward journey at Lo level to evade hostile radars. In the ferry role the new aircraft had to reach Europe without air refuelling, or South-East Asia with one, and this was construed as a ferry range of 3,000 nautical miles. No figures even remotely like these had ever been associated with a tactical aircraft; they certainly called the meaning of the word 'tactical' into question.

The requirement of the Navy appeared to be somewhat different. The Fleet Air Defense (FAD) mission called for an all-weather fighter (strictly, an interceptor) that was compatible with an aircraft carrier, and which could loiter for the longest possible endurance on Combat Air Patrol (CAP) whilst carrying a heavy load of radar and air-to-air missiles. In 1958 the analysts studying the problem had come up with the novel idea of putting much more of the burden of effecting the interception on to the missile. If the missile could fly far enough and fast enough, the carrier aircraft was of less importance; instead of being a supersonic fighter it might be more like a slow transport or utility machine. Studies were made showing extremely large (1,000lb) high-performance missiles carried by such pedestrian machines as the Grumman S-2 Tracker, a 280mph carrier-based anti-submarine aircraft. Rebuilt with powerful search, tracking and missile-guidance radar, the S-2 appeared capable of serving as the air-defence fighter to guard the fleet! By 1959 Douglas had a study contract for a specially designed 'fighter' not very much faster than the S-2 (about Mach 0.8) called F6D-1 Missileer. Its job was to carry a large and powerful radar and computerised air-interception system being developed by Hughes, and a clutch of monster air-to-air missiles called Eagles. This was the exact opposite of the excellent F-4 Phantom then about to come into service; this flew at Mach 2.5 but carried only the relatively short-ranged Sparrow missile, and the even shorter-ranged Sidewinder.

Though the figures kept confirming that the F6D, with exceptional flight endurance and missiles of unprecedented flight performance, could meet the Navy's need and provide sure-fire stand-off defence against hostile aircraft and possibly even hostile missiles, those who had to answer for it to Congress had a nagging fear that a 'slow fighter' was crazy. Finally, in December 1960, the outgoing SecDef, Thomas S. Gates Jr, cancelled the F6D outright, though he kept development going on the radar interception system and Eagle missile.

This was the situation when, a month later, McNamara was briefed on current programmes. Foremost in his mind was the concept of 'commonality' in defence procurement. (Commonality had been proposed by a DoD office several years before McNamara's arrival, but he was the SecDef who really pushed it.) He had not been alone in thinking there was far too much diversification between and within the armed forces. With the simplest items, such as hand hammers, he reckoned to cut the number of types by several hundred, thereby saving a few thousand dollars. With supersonic aircraft he considered the savings of cutting out a single unnecessary type might be of the order of millions of dollars. To his chagrin there were not many areas where he could make such savings, but the SOR-183 and FAD fighter interested him exceedingly. Both were viewed as projects of exceptional magnitude. The Navy aircraft was regarded as the successor to the F-8 and F-4 as the standard defence fighter of every carrier and probably the Marine Corps. The new TAC aircraft, though planned as the replacement for the F-105, was in fact also going to replace the F-100, F-101, F-104, F-110 (later F-4C) Phantom, and specialised attack aircraft, leading to a likely production run of more than 1,500 for the Air Force and Navy plus another 1,000 to be bought by Allies. Within a week, McNamara had decided that, if it could be done without contravening the Earth's natural laws, he would meld together these two gigantic programmes so that the dollar savings would forever serve as a shining example of how things ought to be done.

He called in the Top Brass from the Air Force and Navy, talked with his Director of Defense Research & Engineering (Dr Harold S. Brown) and set a small team of 'Whiz Kids' at work studying the level of cost savings that might actually be achieved. In his own words, 'I saw that these two planes would have many common missions and require many similar operational capabilities. After consultation with my military and civilian advisers, and independent study, I became convinced that one tactical fighter could be developed that would meet both the Navy and Air Force requirements. Accordingly, I directed that the Air Force re-orient its program, with Navy participation, to achieve the goal of a common tactical fighter'.

The prospect looked positively glittering. At one stroke the new SecDef considered he could provide the Air Force and Navy with the aircraft they needed and save a sum his 'Whiz Kids' put at a billion dollars.

The TFX Competition

Usually Americans excel at identifying new concepts and giving them names. This is extremely valuable; giving a new thing a name almost brings it into identifiable existence, and makes discussion faster and unambiguous. But with this new combat aircraft, which was expected to be the most important of the entire post-World War 2 era, the naming business went slightly off the rails; and this influenced the course of the programme. SOR-183 had described the requirement as a 'Tactical Strike Fighter'. During the 1950s the concept of a fighter had become blurred. It used to mean, unequivocally, an aeroplane whose purpose was to destroy hostile aircraft in the air. In the closing months of World War 2 the Allied fighters in Europe had been able to spend nearly all their time strafing and bombing ground targets, and this was also the chief 'fighter' mission in the Korean War in 1950-53. TAC had increasingly become oriented towards air-to-ground operations, and this was further accentuated by the fact that the air defence of the United States was vested in a different command, today called Aerospace Defense Command (ADC). So TAC's formidable order of battle comprised aeroplanes called fighters, and designated with F-series numbers, but which were primarily attack aircraft, sent to clobber surface targets with bombs, rockets and air-to-surface missiles.

Originally the F-100 had been a fighter, but early in its life it was modified to carry heavy loads of air/ground ordnance and air/ground navigation and weapon-aiming systems. The F-101 had been a long-range escort but it was likewise modified for tactical attack and reconnaissance. The much smaller F-104 had been planned as a true fighter, but it was again

Right: The one that got away, two views of a tunnel model of the Boeing-Wichita proposal that most observers thought would win the TFX contract. This model was tested at Wichita State University. / *Jim Rotramel via Jay Miller*

turned into a tactical strike machine and later redesigned specifically for this purpose with mission equipment that almost ignored air-to-air combat. The F-105, which the new aeroplane was specifically going to replace, had never been a fighter at all but was planned as a delivery system of tactical nuclear weapons. So what exactly did 'Tactical Strike Fighter' mean? It meant an attack aircraft, capable of penetrating hostile airspace to deliver to surface targets nuclear weapons or heavy loads of conventional bombs and missiles. Of course, it was meant to have some capability in air-to-air combat, and the requirement stipulated an M-61 internal 20mm gun (though this was as much for use against surface targets as against hostile aircraft). But the concept of the true fighter had in 1960 become pushed into the background. TAC's aircraft acted like bombers, while ADC's all-weather interceptors were designed to steer by radar and computers and launch missiles that would destroy distant enemy aircraft never even seen by the American pilot. The old-fashioned dogfight, in which two pilots in eyeball-to-eyeball confrontation pushed their throttle levers through the firewall in an endeavour to get on each other's tails, was thought to have died with the Sabre and MiG-15.

So SOR-183 was polarised around the long-range air-to-surface mission (today called interdiction strike). The requirement gelled just at the time that the Dulles policy of massive retaliation was giving way to flexible response, and this meant that tactical aircraft had to have the ability to carry heavy bomb loads. As an alternative to two nuclear bombs inside an internal bomb bay, the SOR-183 aircraft had to be able to take off whilst festooned with HE almost from wing-tip to wing-tip. But the Navy FAD fighter likewise had to carry a heavy load of the biggest air-to-air missiles ever developed, and McNamara must be forgiven for thinking the two missions at least superficially similar. What made the similarity assume a harder and more unyielding form was that when the discussion on the two aircraft surfaced in Washington in early 1960 the word 'strike' was omitted, and the programme became known as TFX, for Tactical Fighter Experimental. Unfortunately, while it is possible to design a good fighter (such as the modern F-15 or F-16) and then fit it to fly reasonable ground-attack missions, it is not possible to design a long-range interdiction aircraft and turn it into a superior dogfighter.

On the other hand, nobody in 1960 was asking for a dogfighter, even though the air-superiority mission was one of those that the Air Force wanted their new 'fighter' to accomplish. To most people not closely involved with the two projects it really did seem that, with a little give and take on both sides, a single basic type of aircraft ought to be able to fulfil the requirements of both services. It had to be highly supersonic, have two engines, two seats and probably a swing wing, in both cases. So on 14 February 1961 McNamara issued a formal recommendation, via Dr Brown, that the Air Force and Navy should study the development of a single aircraft to be developed in two versions, with maximum 'commonality', to meet all the requirements. He expected there would be a few hidebound old diehards who would be so imbued with inter-service rivalry that they would be unable to view the situation objectively. Everyone else, he thought, would work a little harder to find the best compromise solution. After preliminary meetings had established that there were indeed problems — in fact, many of the Top Brass appeared quite shocked — McNamara progressively dug in his heels. Working largely with his own civilian staff, who by now were being called not 'Whiz Kids' by the military but 'the bow-tie bastards', he decided that he would have to start ordering people what to do. The best course of action appeared to be to base the TFX on the Air Force requirement and then from this derive a modified TFX for the Navy. Detailed analyses were made of optimised Air Force aircraft, optimised Navy aircraft, minimum-change Air Force aircraft to fit most of the Navy requirements, and optimum 'compromise TFX2' proposals somewhere in between.

These studies involved a large staff of analysts and engineers, many of whom were civilian or uniformed personnel of the Air Force and Navy. But neither service had ever agreed to the concept of a 'common' TFX, unless this was achieved by the other service giving up its critical mission requirements. On 22 August 1961, after weeks of detailed analyses of compromise aircraft, the Secretaries of the Air Force (Eugene M. Zuckert) and Navy (Paul Fay) bluntly told McNamara that it was 'not technically feasible' to build a compromise aircraft that would fly the missions of both services. Though both requirements appeared to call for a swing-wing two-seat aircraft with two afterburning turbofan engines, the similarity stopped there. The Air Force wanted a structure stressed to 7.33 *g* for Lo-Lo-Hi missions at about Mach 2.5 (Hi) and 1.2 (Lo), with tandem seating, a slim nose for good attack visibility and a length in the order of 66-72 feet. The Navy wanted a structure stressed to 6 *g* for high-altitude interception at Mach 2.0/1.0 (the low-level performance was secondary), with side-by-side seating, a fat nose to house a 48-inch radar dish and a maximum length of 56 feet. Secretary Fay stressed to McNamara that the Navy estimated the overall effectiveness of the compromise TFX in the CAP mission at a mere 37 per cent.

On 1 September the SecDef, who had not the slightest doubt he was right, flatly rejected this joint submission. He directed that 'A single aircraft for both the Air Force tactical mission and the Navy fleet air

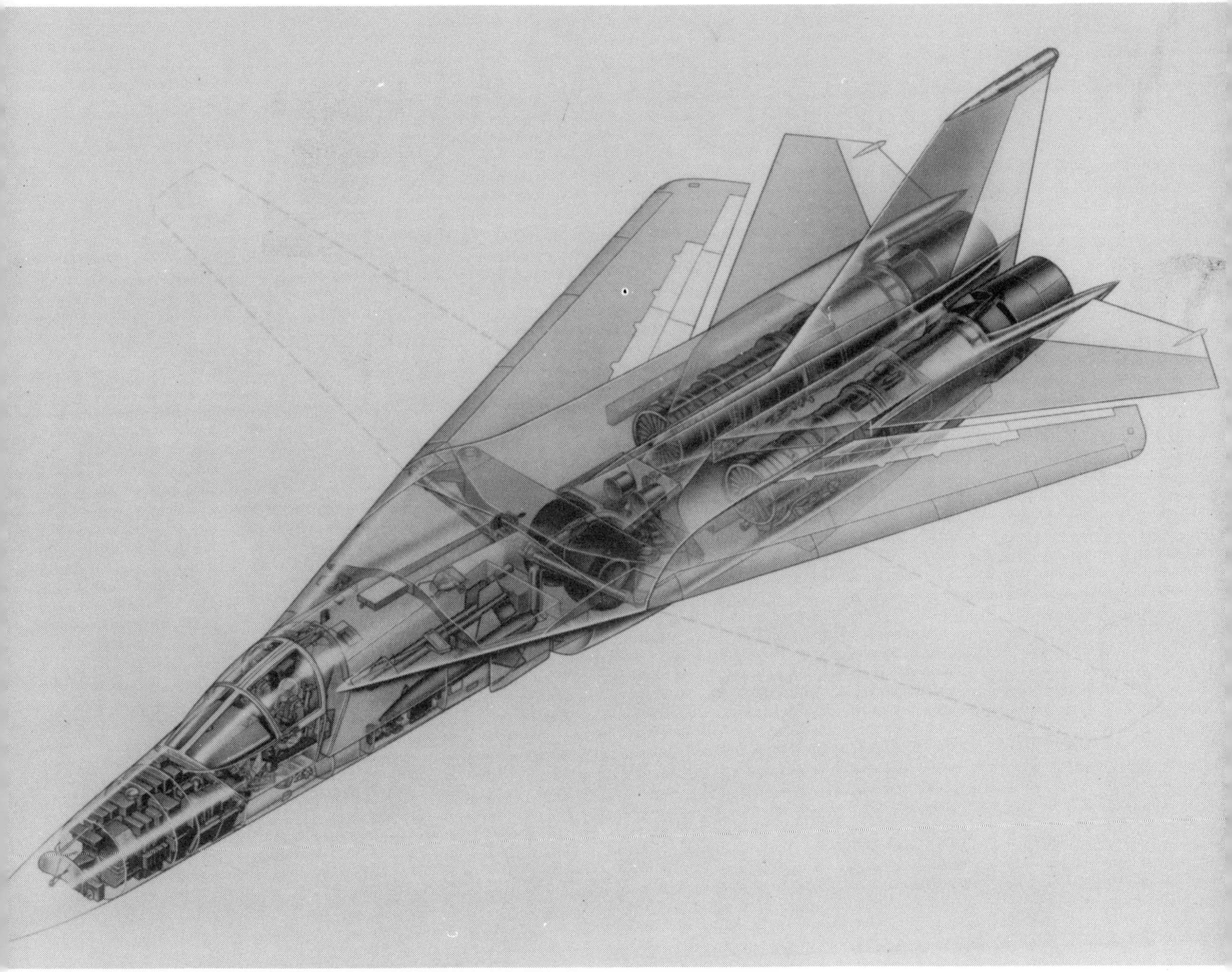

defense mission will be undertaken; the Air Force shall proceed with the development of such an aircraft'. He then stipulated the following design guidelines:

1 The mold line . . . shall be configured so that a radar dish of 36 inches minimum diameter may be accommodated in the nose.
2 The maximum length . . . shall not exceed 73 feet in the Air Force version.
3 The weight . . . in the Air Force version with full internal fuel and 2,000lb internal stores should be approximately 60,000lb.
4 The aircraft shall be capable of delivering a minimum of 10,000lb of conventional ordnance.
5 The design shall allow for the carrying of at least two 1,000lb air-to-air missiles internally or semi-submerged.

Above: Simplified cutaway of the Air Force TFX, at the Fourth-Round submission, showing interior equipment. In the weapon bay are a Bullpup air/surface missile and two Sidewinder air/air missiles, weapons never carried by an F-111 in service. / *GD*

6 The structure must accommodate the loads associated with carrier operations . . . The Navy version . . . shall be capable of performing the FAD mission carrying six 1,000lb missiles at a radius of 150 nm with 3.5hr of loiter time. Takeoff gross weight for this mission shall not exceed 55,000lb without the concurrence of the Navy.'

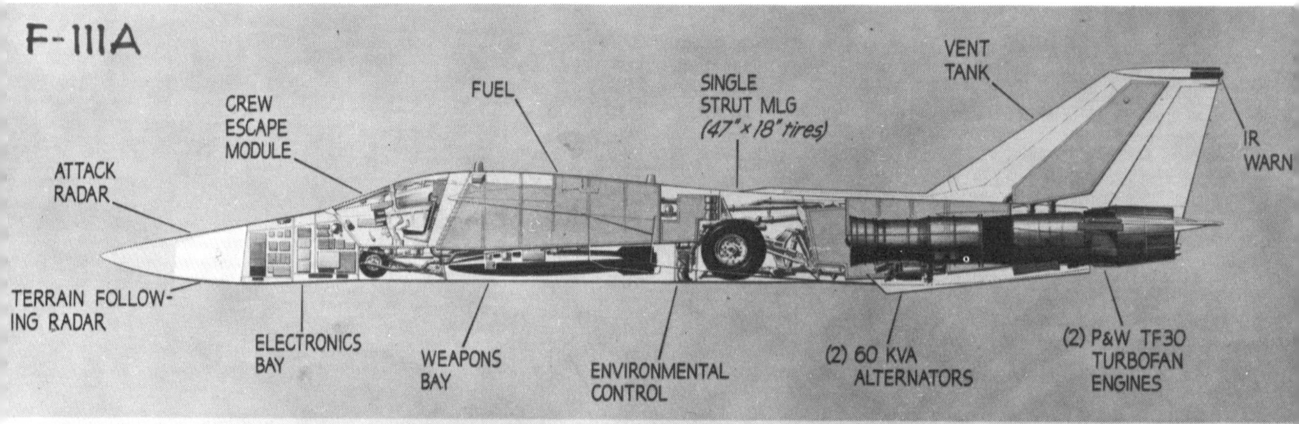

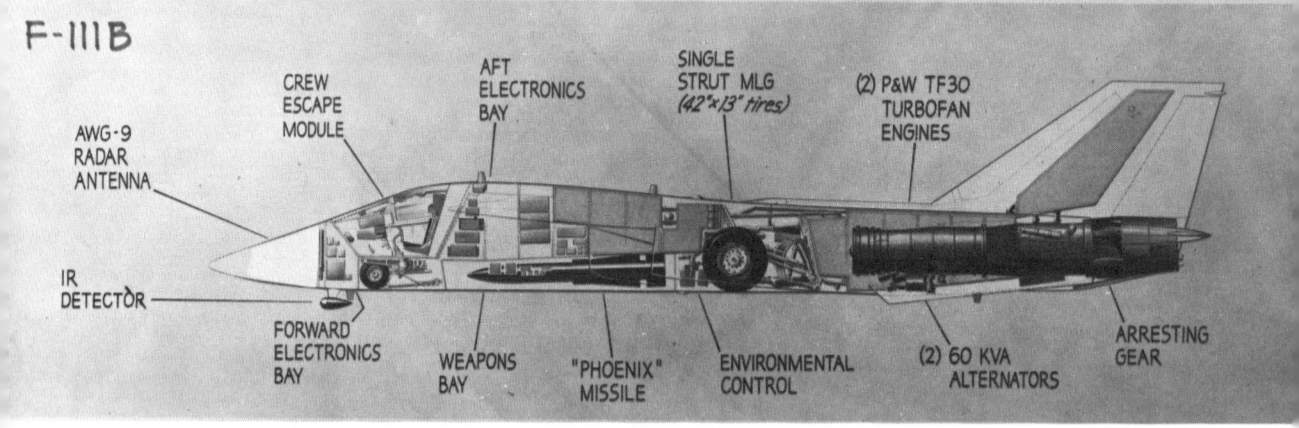

Above: Simplified inboard profiles of the Air Force F-111A and Navy F-111B showing the similarities and differences. The chief difference, apart from the radar/weapon system, was that the Navy aircraft had to be shorter, with avionics in place of fuel behind the crew module. / *GD-Grumman*

The document further stipulated that the RFP (Request for Proposals) should be sent out to industry on 1 October 1961, and, though both the Air Force and Navy agreed the task was impossible, they did as they were ordered and sent out a 'joint TFX Work Statement' on the specified date. For months previously the planemakers had been closely involved with the progress of the controversy, and several had already prepared the bases for compromise TFX submissions. On 6 December six sets of proposals were received from Boeing, Wichita; General Dynamics, Fort Worth (GD); Lockheed, California; McDonnell; North American; and Republic. The immense piles of documents were assessed for six weeks by the 360-strong Air Force Evaluation Group, specially augmented by staff from the Navy and NASA. Their finding was that none was acceptable, but that the submissions receiving highest marks should be refined under paid study contracts. These two submissions were Boeing and GD, but it was already clear that most Air Force evaluators favoured Boeing. On 19 January 1962 the Systems Source Selection Board, comprising representatives of TAC, Systems Command, Logistics Command and the Navy BuWeps (Bureau of Weapons), recommended Boeing. This proposal scored top in flight performance (except for peak Mach number) and weapon carriage, and thus came first in operational capability, but the General Electric MF295 engine was found unacceptable and changes were required in crew escape (a capsule was needed instead of two ejection seats), the radar, and missile stowage.

Boeing was paid to switch to the preferred engine, the Pratt & Whitney JTF10, and both companies made fresh submissions on 2 April 1962. Though many explicitly notified deficiencies had been rectified, the Navy was unable to accept either proposal because neither was fully compatible with carrier operation nor offered adequate CAP loiter endurance. Most Air Force feeling had hardened in favour of Boeing, but, after discussion with the new Navy Secretary, Fred Korth, the decision was taken to request the SecDef for a third 'short round' evaluation taking three weeks to allow final correction of deficiencies. Yet, when the third set of proposals were in, with almost unanimous voting for Boeing by the Selection Board, Air Force Council and Chief of Naval Operations, the result was a call for an unprecedented fourth round. This was launched on 29 June 1962 and the proposals were received on 10 September. Again, the Selection Board, the commanding generals of the Systems and Logistics Commands, the Chief of the Navy BuWeps, and General Sweeney, the new Commander of TAC, all recommended selection of Boeing. In its final document, dated 8 November 1962, the Air Force Council stated 'There is a clear and substantial advantage in the Boeing proposal over the General Dynamics proposal, which is magnified by the environment found under the austere conditions usually inherent in limited-war actions'. It was therefore a considerable shock when, on 24 November, the Defense Department issued a public press release announcing the award of the TFX to General Dynamics.

There was an immediate outcry, and later the chairman of the US Senate Committee on Government Operations, Senator John L. McClellan, held a long and much-publicised series of hearings at which the manifold suggestions of malpractice, if not outright fraud and scandal, were given a thorough airing. Not until he was called as a witness at these hearings on 21 March 1963 did McNamara say why he had chosen the GD proposal. In his own words:

'The report (of the fourth evaluation) evaluated General Dynamics as having a better structural design, a simpler fuel system, a slight edge in the flight-control area and better proposed programmes in the personnel subsystem and aerospace ground equipment areas. The GD design had an edge in supersonic dash capability and supersonic manoeuvrability at altitude. It has a low radar cross-section and an integrated penetration-aids system. For deceleration it used dive brakes in the air and brakes on the ground . . . The Boeing design has the edge in ferry capability, conventional-weapon carriage, loiter capability and landing performance. It has the advantage in low-altitude manoeuvring capability. For deceleration it uses a thrust-reverser which offers an excellent performance but will require additional development effort.

'When I reviewed the report I was immediately struck by the difference in approach adopted by the two contractors. The report found that GD proposed an airframe that has a very high degree of identical structure for the Air Force and Navy versions. In the two Boeing versions less than half of the structural components of the wing, fuselage and tail were the same. In fact the evaluation group concluded that Boeing is, in effect, proposing two different airplanes . . .'

Despite this, Boeing's bid was priced lower. Contractors had been asked to bid on the basis of a fixed price for 22 research, development, test and evaluation (RDT&E) aircraft, five of them of the Navy version, and to estimate production cost on a further 1,704, of which 231 were the Navy version. Boeing's fourth-round price was $466.6m for RDT&E plus $4,897.7m for production, a total of $5,364.3m. The GD price was $543.3m plus $4,912m, a total of $5,455.5m. But McNamara and his 'Whiz Kids' unearthed what they called 'evidence of unrealistic cost estimates in the Boeing proposal', and also criticised Boeing's engine installation, with dorsal air inlets, thrust reversers, titanium primary structure and several other design features. In a nutshell, the SecDef was saying he did not believe Boeing could do what they offered, and certainly not at the price. In fact, there were at least four sets of cost figures: those above, the same figures adjusted to common standards by the Air Force, Air Force Cost Standards, and the last-named after correction for error. In all these Boeing's figures were significantly ($91m to $416m) lower than those for GD, but the SecDef interpreted this as 'greater cost realism on the part of GD'.

Senator McClellan's TFX contract investigation remains a classic probe into defence procurement. Coming at the same time as an almost equally monumental scandal revolving around the Lockheed C-5A logistic transport, which was greatly down in capability, up in price and suffering from severe technical difficulties, it cast a great cloud over the procurement policies of the Pentagon just at the time the boldest and most dynamic SecDef for many years was trying to improve matters. There was nothing the powerful Senator could do to reverse the TFX contract award, though efforts had started even before the actual contract had been signed on 21 December 1962. We will never know if Boeing would indeed have produced a better or cheaper aircraft. But we can be fairly sure that their development difficulties would not have been as great as those that hit General Dynamics.

GD hits Turbulence

When it won what the McClellan investigation called 'the largest single airplane contract that had ever been awarded up to that time' the Fort Worth division of GD was muted in its celebration. The gruelling four rounds of evaluation had consistently left them in second place, and the sudden announcement of the contract award — followed so closely by outcries and denunciations — left a nasty taste in the mouth which could best be erased by turning in a superior contract performance. GD had bid in partnership with Grumman, a most experienced contractor for Navy aircraft, who were charged with design and production of the aft fuselage and landing gear and the development, assembly and testing of the Navy version. The latter was now designated F-111B, the Air Force version being the F-111A. It was the last 'fighter' to have a Century-series number, because in 1962 the Pentagon introduced completely new numbering in which Air Force and Navy aircraft have a uniform system starting again at No 1 in every category.

Today's F-111 has a familiar shape but, like a Beethoven symphony, it got that way only after a long process of alteration and refinement. Almost the only assumptions GD and Grumman made were that the crew compartment should be near the front, the tail at the back, the engines side by side (and in most studies well separated laterally) and the wing fairly high to get above the weapon bay and inlet ducts. Project studies had started in early 1961, but when the RFP went out on 1 October of that year the Fort Worth tunnel models were still coming in all shapes, with fat and slim bodies, pointed or stubby swing wings, various

Right: The original YF-111A (63-9768) is outwardly almost indistinguishable from the inventory aircraft of today, apart from its grey paint scheme; but in fact there are hardly two parts common. Note tails of B-52s at Carswell AFB, Fort Worth. / *GD*

U.S. AIR FORCE

Above and right: Two views looking down on the first YF-111A early in its test programme soon after Johnson had first demonstrated that it was safe to exercise the swing wings over their whole range of sweep from 16° to 72.5°. / *GD*

Bottom right: Close air-to-air of the first development aircraft at minimum sweep showing the original fuselage base which was later redesigned. / *GD*

nose shapes and with tails high, middle or low. A photograph shows some of the more likely forms on 30 November 1961, a week before the submitted design (which was not quite like any of those shown). It will be especially noted that all featured supersonic inlets projecting far ahead of the fixed wing-glove. Another picture shows the original submission aircraft. This has subtle curves, long and straight engine ducts, a low tail and rectangular blunt base to the rear of the body.

Two further photographs show the second-submission aircraft, and in this the major changes to try to meet requirements have changed the shape almost to that of today's aircraft. The nose has been made shorter, with the side-by-side seats established in both versions. The fuselage base has been tapered off in a narrow fairing, and the vertical tail moved forward alongside the bulged engine nozzles. The top of the wing and fuselage has been made almost flat, giving extra lift, and the most significant change of all is that the long inlet ducts have been cut right back so that the inlets are under the wing, the gloves being extended forward to a point just behind the canopy, portions of glove serving to help stabilise the ejected crew-module.

Above: November 1965 photo showing the new Fort Worth F-111A flight-line ready hangar and three development prototypes. At that time eight aircraft had logged 397 hours in 282 flights, more than scheduled despite the engine-stall problem. / *GD*

Of course, even at this early stage the F-111A and B each suffered considerably from the conflicting needs of the other. Both could have been greatly improved if only McNamara had relented in his burning desire to 'save a billion dollars' by the concept of *maximum 'commonality'*. One of the unchangeable features that displeased the Air Force was that pilot and navigator sat side-by-side, and if there is one thing that bothers a tactical pilot (especially one flying a supposed 'fighter') it is having no better view than a transport pilot, being able to see ahead and to the left but not to the right or rear. GD and Grumman actually worked wonders in ending up with an Air Force aircraft with a length of 75ft 6in and a Navy one held to 66ft 9in. Structurally their commonality was high, as McNamara had stressed, the main differences being the different lengths of fuselage (neatly accomplished entirely ahead of the cockpit) and the fact the Navy aircraft had slightly longer swing wings. Other differences, such as tyre sizes, were relatively minor. Fuel system, engines, weapon bay and primary structure were almost identical.

This was the central reason why GD/Grumman won the programme. But the extreme pressures on the design team to maximise commonality had led to serious technical problems that were to exact terrible penalties.

When GD/Grumman were named winners on 24 November 1962 they contracted to fly the first F-111A in 25 months. This meant moving fast, and because of the size of the programme it needed a lot of manpower even by American standards. Total wind-tunnel test time at first flight was no less than 20,336 hours (and, quite unpredictably, it eventually more than doubled, because of the problems encountered) and NASA's Langley Lab did a lot of tunnel testing under subcontract to GD. Though Stack was by this time generally called 'Mr TFX' and had every reason to wish the programme to go well, his aerodynamicists soon found the shape was still not right. In March 1963 they discovered 'major supersonic problems in the form of high drag, low manoeuvrability, low directional stability, and a serious transonic drag problem'. As 1963 progressed, the weight engineers at Fort Worth likewise reported deep trouble. Though the estimates of equipped empty weight had been refined in four rounds of detailed bids, the final F-111A figure being 36,700lb, it was clear by mid-1963 that the actual weight was going to be considerably higher.

This was as serious as the shortfall in performance due to excessive drag. It soon looked as if the gross weight with conventional bombs was going to reach 70,000lb, and almost 60,000lb for the Navy F-111B. It is not true to suggest, as many observers have done, that the Air Force positively wanted a long and heavy aircraft. At the outset TAC optimistically hoped the gross weight could be held to 45,000lb, and at a 1963 programme review meeting at Langley AFB an aggravated Gen Everest insisted 'I'm not going to accept any goddamned 70,000lb airplane'. But with too much weight and too much drag, the prospects for meeting the mission performance, especially the Lo-Lo-Hi attack radius, were almost non-existent. The

Above: Swing wings are usually spread on the ground, but in this picture the first development aircraft is taxiing at an angle close to maximum sweep. Note the large flat plate used on the original inlet to divert boundary layer. / *GD*

Centre left: The fifth aircraft is seen here carrying dummy B61 nuclear weapons on its four swivelling pylons. Even the first aircraft had provision for pylons but these were not fitted. It was soon discovered that the swivelling mechanism was prone to failure. / *GD*

Bottom left: Fifth aircraft above cloud during early tests with pylon loads, in this case practice weapons. The largest external store, the 500gal (600 US gal) tank, once hit the fuselage violently after release. / *GD*

Right: View from above the fifth development aircraft, for comparison with similar photographs of No 1. Pylons are fitted and the limit of the fuselage walkway area increased. / *GD*

main hopes lay in superior engine performance, and in detailed engineering changes with the RDT&E aircraft to shave off some of the excess weight.

Pratt & Whitney had a proud tradition of sound and conservative engineering, leading to dependable performance. Their JTF10 two-shaft turbofan had begun life in 1958 as a subsonic civil engine for a four-engined short-haul airliner called the DC-9 that Douglas never built (it bore no relationship to today's DC-9). For the TFX programme this engine was restressed for Mach 1.2 at sea level, fitted with a large afterburner and variable nozzle and equipped for military use, designated JTF10A-20. Later it received the service designation TF30-P-1, and in its early bench trials in the summer of 1964 it behaved satisfactorily, the completely new feature of afterburning in the fan airflow giving little trouble at sea-level pressures. Even simulated high-altitude testing in the Willgoos Lab was encouraging.

By this time the RDT&E batch of aircraft had been increased to 23, and the first to be completed, YF-111A serial 63-9768, was ceremonially rolled out 16 days ahead of schedule on 16 October 1964. After extensive ground testing it took off on its maiden flight, still 10 days early, on 21 December, in the hands of Dick Johnson, GD chief of flight test, with Val Prahl as observer. Those watching were elated. As the long-nosed ship rotated and climbed away, the bulging balloon tyres, sized to permit operation from unpaved airstrips, sagged lower and lower as the long-travel main gears reached full extension. They were kept down on this first short workout, with the wing locked at the 26° setting. For the second flight, on 6 January 1965, the wing panels were unlocked. Johnson was hopeful that he could earn GD the maximum cash bonus of $875,000 by demonstrating the full range of variable sweep 24 days ahead of schedule. After cautiously inching the swing wings forward to the minimum 16° setting he gradually worked his way back, until eventually the wings were folded right back at the maximum 72.5° setting. In this position the F-111 should easily go supersonic, and Johnson cautiously applied more power in level flight. Demonstrating supersonic flight on Flight 2 instead of the scheduled Flight 3 would earn a further bonus. It led instead to unexpected trouble; there was an almighty hiccup and a noise that Johnson knew only too well from other aircraft. The engines had suffered severe compressor stall.

GD knew they were running a risk of this when they cut back the length of the engine inlet ducts. Inside the engine are rows of large fan blades and small

compressor blades, revolving at high speed and behaving like miniature wings. If a single blade encounters air hitting it at the wrong angle (called angle of attack) it can stall, causing complete breakdown of the airflow, sudden turbulence and, in especially severe cases, breaking off the blades. The engine designer tries to provide a margin between the engine operating conditions and the stall line, the boundary of conditions at which it will stall if it encounters any turbulent air. But a bigger stall margin means a lower engine performance, and usually higher fuel consumption. To meet the demands, Pratt & Whitney had hewed too close to the stall line. Trouble might still have been avoided if the F-111 had retained its original long inlet ducts, which might have smoothed out the airflow sufficiently. Of course, for combat aircraft the engines must give full power even in violent manoeuvres, when the air may enter the inlets obliquely from almost any direction, causing breakaway and turbulence inside the duct. With the very short inlet ducts chosen for the F-111 the problem was terrible. On Flight 2 the engines stalled in straight and level flight, when the airflow ought to have been smooth.

Flight testing had to be suspended on 5 February 1965 while GD investigated 'erratic engine behaviour'. Pratt & Whitney urgently produced a considerably modified engine, the TF30-P-3, which among many other changes had different blade angles and rotational speeds and a 6th-stage bleed to increase the stall margin at high Mach numbers or in combat manoeuvres. Partially modified engines incorporating all the anti-stall measures were rushed to Fort Worth by mid-February. Johnson resumed flight testing and suffered compressor stalls as bad as ever. The GD team working on the inlet ducts was at once augmented and, over a period of several months,

Top left: Sixth development aircraft pictured during flight-refuelling trials in 1965. Rudder and stabilator tip are painted white. / *GD*

Above: Unusual picture of one of the YF-111A development aircraft at 16° sweep with eight pylons each carrying three 500lb bombs. The outer four pylons do not swivel and have to be emptied and jettisoned in order to exceed Mach 1; in service they have seldom been used. / *GD*

Top right: Aircraft No 10 at 26° sweep in 1966. By this time most of the original problems had surfaced. / *GD*

achieved an inlet and duct system that allowed the P-3 engine to avoid stalls over most of the flight envelope. The inability of the F-111A to manoeuvre as tightly as the Air Force had demanded was a positive help at this point, though not to the Air Force. Called the Triple Plow I inlet because of its three sets of lips to divert the sluggish boundary-layer air, and with vortex generators sprinkled over its walls, the new pattern was finally cleared in early 1966, by which time most of the 18 development F-111As had flown.

Supersonic speed was not reached until Flight 9, on 5 March 1965, with an interim inlet system and engines not yet at P-3 standard. In his superficially damning report Senator McClellan was scathingly critical of the refusal of the SecDef and GD to stop F-111A production in 1965 and get the aircraft right. As it happened, there were plenty of major problems still to surface, but nobody knew this at the time. What was known was that the P-3 engine, rated at a maximum thrust of only 18,500lb, was not powerful enough for an F-111 whose weights were escalating in an almost uncontrolled manner, and that even with the Triple Plow I inlet the engine installation was on the limits of the airflow it could handle. There is no doubt that, just as Pratts pushed their luck with the original TF30 compressor, GD cut things too fine with the small and cut-back inlet system. Later in 1966 a more radically modified Triple Plow II inlet was developed, described later. This allowed Pratt & Whitney to substitute more powerful versions of the TF30 as these became available, as described in later chapters.

Superimposed on the severe problems with the basic aircraft was a major development programme to produce the RF-111A reconnaissance aircraft. Such a development was obvious, because the F-111 was an outstanding basic platform for such work, and from early in the TFX studies it had been decided that the weapon bay should be able to carry bombs, or a gun, or a multi-sensor reconnaissance pallet depending on immediate tactical requirements, with less than an hour needed for the changeover. In practice, of course, an RF proved to be more than a mere clip-on pallet. McNamara announced the project in December 1965, and the 11th aircraft was completed with the necessary basic modifications for conversion. But it did not fly until 17 December 1967, a cool $118m later, with a complex digital control system governing cameras, SLAR and IR linescan sensors in a pod filling the weapon bay, with aerials and apertures in a fairing larger than that of the gun. And that was the last we ever heard of it, and today the USAF and Pentagon profess to have 'nothing on file'. Was it a purely political decision to kill this potentially important aircraft? What makes the joke even more sardonic is that millions have since been spent trying to develop a reconnaissance pallet for the Australians, as described later.

One of the basic advantages of the TFX programme was that it was supposed to yield a massive production run of at least 1,726 aircraft for US customers and possibly 1,000 more for Allied and friendly air forces. The expected benefits in reduced unit price were large. A 1962 estimate for the price of two TF30 engines was a mere $540,000, and this was based on a modest run of 2,000 identical engines, considerably fewer than the number everyone thought would be built. But what actually happened was that both the aircraft and the engine were continually being modified, while the numbers of aircraft in the overall programme dropped and dropped. Thus the benefits of long production runs vanished. So, too, did the 'billion dollar saving' of 'commonality' with the Navy, as explained in the next chapter.

Exit the Navy

Putting the F-111B together was mainly the problem of Grumman and Hughes, the latter having responsibility for the radar/missile system that the aircraft existed to carry. Despite its impressive performance, the Eagle missile had been cancelled for political reasons during the 1962 discussions on 'commonality'. This was over the heads of the admirals, who never ceased to insist that the whole purpose of the Navy TFX rested on its multi-target stand-off interception capability, which in turn rested squarely on the powerful radar and missile. After much further discussion, centred around a cogently argued Navy document called *Fighter Study I,* a new air-to-air system was permitted to arise from the ashes of the old. Hughes was awarded major contracts for the modified F6D radar intercept system, now called AWG-9, and for an even better missile called AIM-54A Phoenix. The Navy relaxed very slightly and, though still deeply suspicious of both 'commonality' and the Air Force-based way in which it was being applied, trusted in its capable and experienced contractor Grumman to see it through.

It was inherent in the concept of 'commonality' that, though the Navy wanted a smaller and lighter aircraft than the limits acceptable to the Air Force, the special requirements of the carrier-based interception mission greatly increased the weight of the tactical F-111A version that was the starting point for the programme. Then, in working in the reverse direction to create the Navy F-111B version, with 'minimum change', the penalties of 'commonality' became even greater. From the start GD and Grumman were aware that the weight problem with the F-111B would be extremely serious. Though the first three development B-models had to be built and flown, a Super Weight

Right: Ceremonial roll-out of the YF-111B at Grumman's Peconic River plant on eastern Long Island on 11 May 1965. Everyone hoped it would be a worthy stablemate to the famous Grumman aircraft that surrounded it.
/ *Grumman*

Grumman

Above: Accompanied by an A-6A Intruder, the first YF-111B cruises serenely over Little Peconic Bay on its first air-to-air photographic sortie in May 1965. / *Grumman*

Right: Another air-to-air of the first YF-111B off Long Island, at minimum sweep and with the airbrake (landing-gear door) open. This brake caused buffet problems. / *Grumman*

Below: The second YF-111B at intermediate sweep; note the open auxiliary air inlet around the engine duct. / *Grumman*

Above: Probably the F-111B was the only F-111 that might be called pretty, especially at 72.5° sweep. / *Grumman*

Improvement Program (SWIP) was launched to try and cut every possible pound out of aircraft No 4. Meanwhile, according to McClellan's probe, a 'policy of concealment' was followed to try to prevent the true extent of the weight problem surfacing, in the hope that SWIP might cure the worst of it and thus avoid a major public row. The monthly weight statements of estimated equipped empty weight of the F-111B rose from 38,804lb, in the final September 1962 fourth-round proposal, to 39,459lb in February 1963, 39,861 in March, 39,864 in July, 40,276 in September and 40,284 in November. But in December a quite different source reported the same weight as 45,259lb, and in July 1964 the Navy estimate was 46,284, quite close to the originally hoped-for gross weight in the two-missile CAP mission.

So serious were the problems of weight growth, poor aerodynamics and grossly reduced mission effectiveness with the F-111B, even compared with the 1962 proposal aircraft (which was itself in many respects severely compromised by being based on the F-111A) that in January 1964 the Navy began a complete re-evaluation of the programme. The result was a detailed report dated 3 February 1964 which said, in effect, that the F-111B could not go on without 'major changes' if it was to be an acceptable weapon system. Among numerous shortfalls in performance were increases in wind-over-deck speeds for both take-off and recovery that were beyond the speeds available (see table):

F-111B performance (selected figures)

	Sept 62	Feb 64	July 64	July 67
Weight empty (lb)	38,804	45,000	46,284	46,112
Takeoff weight (lb)	62,788	77,000	77,692	79,002
Combat ceiling (ft)	55,000	52,000	50,750	44,000
Single-engine climb (ft/min)	595	270	190	100-
Mach 0.8/1.8 time (min)	5.5	6.2	9.8	10.0+
Launch wind over deck (kt)	-8	+14	+20	+19
Recovery wind over deck (kt)	+5	+21	+22	+10
Aproach speed (kt)	113	124	125	?

On 11 February the Chief of BuWeps, Admiral Masterson, asked that manufacture of the F-111B development aircraft should be halted pending gross redesign, but this was overruled in the Pentagon and the programme continued. In fact, though redesign was obviously necessary, little purpose would have been served by merely halting the manufacturing programme. The first F-111B, serial 151970, flew at Grumman's Calverton field on 18 May 1965, and even at this late date was soon in trouble with compressor stalls. While flight testing proceeded on a fleet of F-111B aircraft that ultimately numbered seven, all different and from No 4 introducing extensive weight-reduction measures — which in practice did little better than counteract natural weight-growth — plans were made and re-made for a Colossal Weight Improvement Program (CWIP) and variations of this called CWIP I to III. What was called 'major surgery', almost wholly in the fuselage, could cut 4,000lb off the structure weight and result in an aircraft marginally capable of flying the Navy's missions. Commonality fell in the process from 80 to 29 per cent.

The first NPE (Navy Preliminary Evaluation) on the F-111B was held in October 1965. It described the aircraft as 'grossly underpowered', quite apart from the fact that the engine stalls were judged 'a safety of flight problem', while range deficiency averaged 44 per

cent. By this time GD/Grumman's main line of attack was to add fuel and fit a more powerful high-lift system of slats and flaps to try to counter the further increase in weight (much the same was being done with the F-111A). While the major surgery picked off some of the fat, the internal fuel capacity was increased from 16,000lb to 26,000lb, putting the takeoff weight close to 80,000lb, or 60 per cent over requirement and 26 per cent above the final proposal.

Pratt & Whitney redoubled their efforts to produce a stall-free TF30 engine with greater thrust, and in 1966 produced a lighter and more powerful version for the F-111B designated TF30-P-12. Rated at 20,250lb with full afterburner, this helped fight the weight growth, as did the significantly higher maximum lift coefficient of the new slat and flap system. But from January 1964 it became clear to a growing number of people that the B-model was so compromised that it was mistaken to continue. Put another way, the 'major surgery' suggested in 1964 was already so major that it became a new aircraft, and by 1966 even this could be seen to be inadequate. Back in 1964 the Navy report had commented 'If the pilots don't like it, it will never reach the Fleet'. Despite its obvious long-range interception capability with the AWG-9/Phoenix system, the basic F-111B aircraft was almost universally regarded as unacceptable. In almost every aspect of performance it was inferior to the F-4 Phantom.

Left: Striking underside view of F-111B No 2 at maximum sweep. On the F-111A the wing tips do not reach to the tips of the stabilator (tailplane). / *Grumman*

Below: This view of F-111B No 2 clearly shows the way in which the engine installations converge. / *Grumman*

Best-looking of all One-Elevens, the F-111B navy fighter finally ground to a halt in mid-1968./*Grumman*

From the start the Navy had been extremely unhappy about the F-111B, and it said so in public — to the delight of the media, and to Senator McClellan. The inescapable gravity of the situation caused McNamara to do something quite extraordinary, and almost ridiculous: he decided to try to run the programme himself. On 25 August 1966 at nine in the morning a galaxy of talent gathered in the Pentagon to solve the problems. It was the first regular Saturday-morning meeting which the SecDef said would continue 'until either the meetings are doing no good or the problems are solved'. Nobody thought to invite Gen A. L. Esposito, the F-111 Project Officer in the Pentagon and the man who knew most about the programme in the whole DoD (in fact, anyone in uniform was not welcome). Someone in Esposito's office called the meetings Project Icarus, perhaps because Icarus had wings which, softened by the sun's heat, suffered gross variable geometry causing him to fall like a stone. GD Fort Worth President Frank Davis acidly commented 'I don't have time left between the meetings to solve the problems'.

In March-April 1967 the Navy ran a completely fresh NPE on the SWIP aircraft, Nos 4 and 5. It wanted to believe that this might be a major step along the road to an acceptable aircraft, but the list of 'correction mandatory' deficiencies was as long as ever. Of course, any NPE throws up many snags; those for the F-4 and F-14 could, if reported in a selective way, be made to indicate that those aircraft were unmitigated disasters. But the F-111B hardly had anything right. Little purpose would be served by repeating even some of the worst faults, but it is worth noting that on paper the F-111B was being almost completely redesigned. Called Navy II F-111B, the result was more than 'major surgery' and virtually a new aircraft. The fuselage was completely different, with a long nose housing the AWG-9 boxes previously behind the cockpit, giving better accessibility and another 2,000lb of internal fuel. The weight distribution was also improved, and the tip-back problem on carrier decks further alleviated by moving the main gears further back. Body pylons were added to replace the swing-wing swivelling Phoenix pylons which were complicated, heavy and sometimes failed to work. A completely new canopy and windshield was fitted which, together with increased flap deflection and a raised pilot seat, improved view during carrier landing. There were many other gross changes and minor fixes.

In mid-1967 the civilian Secretaries in the Pentagon were still putting a bold face on the programme, but in the wings Grumman was using its own money to build a completely new wing box of welded titanium for a different and much better fighter designed to fly the Navy mission and nothing else. The company found no reluctance on the part of the admirals to talk about a new and uncompromised fighter using the TF30-P-12 engine, AWG-9 radar fire control and Phoenix missile, and by September 1967 this had openly become the VFX (Navy code for a fixed-wing fighter, experimental). The following month Grumman also

Left: F-111B No 2 at 26° sweep. The white portions of wing glove form part of the escape capsule. / *Grumman*

Right: An F-111B development prototype with wings at almost full sweep whilst carrying four early XAIM-54A Phoenix air/air missiles. The F-111B swing-wing pivoted pylons were especially troublesome. Note under-nose IR seeker. / *GD*

Below: An F-111B development aircraft takes off from NAS Point Mugu with two early Phoenix missiles (they are possibly inert aircraft-compatibility rounds). / *US Navy*

Bottom: XAIM-54A Phoenix on left inboard swivelling pylon on an F-111B at Point Mugu, with Navy project pilot. Note high-lift flap and slat. / *GD*

proposed VFX-2, with a new advanced-technology engine. Fighter Study II compared the two VFX proposals with the existing F-111B and the Navy II F-111B, but the margins were so large as to be meaningless. While even VFX-1 could accelerate from Mach 0.8 to 1.8 more than 60 per cent faster than an F-4, the F-111B was 50 to 150 per cent slower, taking 6 to 10 minutes. The same order of margin was found everywhere, but it was in what the Navy called the Other Fighter Role, air superiority in close combat, that it was simply no contest. While the F-111B was no kind of dogfighter at all, the VFX promised to outfly all existing fighters. This was the straw that broke the F-111B's back.

In May 1968, despite all that the civilian Secretaries in the Pentagon could do, the Armed Services Committees of both houses of Congress simply refused to authorise funds for F-111B production. That was the end of it. On 10 July 1968 the DoD had to issue a stop-work order, and formal contract termination was agreed with GD and Grumman on 14 December of the same year. The five prototypes and two production Bs had flown 1,748hr in 1,173 flights, and laid valuable groundwork for the engine, radar and missile which were packaged into a new airframe that became the Grumman F-14 Tomcat. Perhaps it is rather unhelpful to suggest that the F-111B succeeded merely in consuming $377.7m, degrading the Air Force F-111 versions and delaying the F-14 six years.

NB
060

Top left: FB-111A trials aircraft for SRAM Cat III testing at Eglin, with four missiles aboard and launch cameras under the tips of the swing wings, under the nose and behind the main gear bay./*GD*

Bottom left: One of the early Nellis Aardvarks, from the 429th TFS, pictured during training missions in 1968./*USAF*

Below: An F-111E of the 20th TFW as seen by the boomer of a KC-135 somewhere in the European theatre. /*USAF*

Bottom: An FB-111A of SAC with four 500-gal (600 US gal) tanks./*USAF*

What d'ya mean, a Lemon?

Lemon was the acidic epithet first applied in print to the F-111 in 1966. The media were rather slow to cotton on to the fact that, as well as being the focus of an unprecedented storm over who should make it, the F-111 was in considerable trouble technically. As with the Navy model the reasons were complex and interwoven, but the original root causes were higher than predicted drag, excessive weight growth and severe shockstall problems with the engine and inlet system. The escalation in weight was compounded by the need to increase fuel capacity to try to restore as much as possible of the lost range. As neither wing area nor engine thrust were increased (in fact the first engines gave less thrust than predicted) the result was an aircraft only marginally useful as a tactical bomber, whose saving grace was its outstanding radar, navaids, penaids and ability to make first-pass attacks on blind targets — as explained in this chapter. Contrary to what had been planned it could not blast off a short airstrip at gross weight, manoeuvre in any way resembling a fighter nor fly any useful fighter mission at all. But this in no sense meant the F-111A was a lemon.

To an American a lemon is something weak, floppy and useless, such as a racing car that consistently comes in last. So far this book has probably given the impression that an F-111 would either come in last or, more likely, fail to finish. That is because its early years really were little more than a chapter of disasters, outwardly. It is impossible to report what happened without seeming to be beastly to McNamara, or Frank Pace's engineering team, or Pratt & Whitney's engine men or Senator McClellan for being beastly to everyone else. Even as it is, we have yet to encounter about half the problems! But

Right: F-111A of 428th TFS in early workout of TFR flight in the mountains near Nellis, 1970. / *GD*

Above: Departure from Fort Worth of the 106th F-111F, the 562nd and last One-Eleven (Serial 70-0188)./*GD*

Left: This happens to be an A-model lifting off from the Fort Worth plant, but the E, D and F are almost identical from a distance externally./*GD*

Above: One of the first F-111As to reach TAC, still with interim back-end (fuselage base aligned with rudder trailing edge, and pointed ECM aerial fairings). / *GD*

from this point on it must gradually emerge from the morass of difficulties that, sorely troubled and compromised as it was, the F-111A could put wholly new capability into the hands of the TAC aircrew. Reading that GD promised a ferry range of 4,180 miles, whereas the actual first-prototype flight article ran dry at 2,300, or that in the low-level dash at full throttle at Mach 1.2 the limit was a mere 30 miles instead of the promised 210, gives an impression of gross failure that is in fact misleading. Though the shortfalls were large, they were shortfalls against paper promises.Comparisons against TAC's previous equipment, or against other aircraft developed in the same timescale, rather surprisingly suggest that the F-111A is really quite some aircraft.

As this previously critical book is about to look on the bright side, now is the time also to forget technicalities for a while and relate some anecdotes of the flight-test programme. They will bring action into a story that was full of it.

Early flying was in the main marred only by the shock stalls, but a fatality occurred on 19 January 1967 when F-111A No 9, engaged in Cat II testing for Air Force Systems Command, suffered an accident filled with irony. The pilot and co-pilot were snugly settled on a radar approach at Edwards when they noticed that the sink-rate was abnormally high. A quick check showed they had the wing at 26° but had set up the low airspeed appropriate to a 16° setting. Wing sweep is worked by a trombone-action pistol-grip. Chief of flight test Dick Johnson recalls 'The swing wing itself gave not a moment's bother. The one thing we all argued about endlessly was which way to move the handle'. His colleague A. S. 'Doc' Witchell, assistant F-111 program director, thought 'Most people visualise the shape of the wing and feel the control should operate in the same sense.' Others said it should go with the throttle, pushing everything forward to go fast and pulling back to go slow. Here was an F-111 nicely lined up with the runway but at the wrong sweep. The pilot could not remember which way the handle worked, pushed it all the way forward, and had everything happen at once. As the wing was pivoting through the 50° setting, at airspeed for 16°, the aircraft hit the ground hard about a mile short of the runway. The co-pilot, completely uninjured, jumped out and went round the nose to help out the shaken pilot. As he leaned over the sill, helping the pilot undo his connections, hundreds of gallons of fuel in which the co-pilot was standing suddenly ignited. The pilot managed to get out almost unscathed, but the co-pilot was burned to death.

A slightly similar arrival was made by GD test pilot Fred Voorhies, who found himself demonstrating a production F-111A in an Armed Forces Day show at Holloman AFB. At one point he planned to make a minimum-speed fly-by in the dirty configuration (gear, flaps, slats and brake extended) and then quickly clean up, hit afterburner and go into a steep climb whilst sweeping the wings. Unfortunately he let speed get a little low and was soon right on the back of the drag-curve, the condition in which less speed means more drag. The only solution is maximum power, at once,

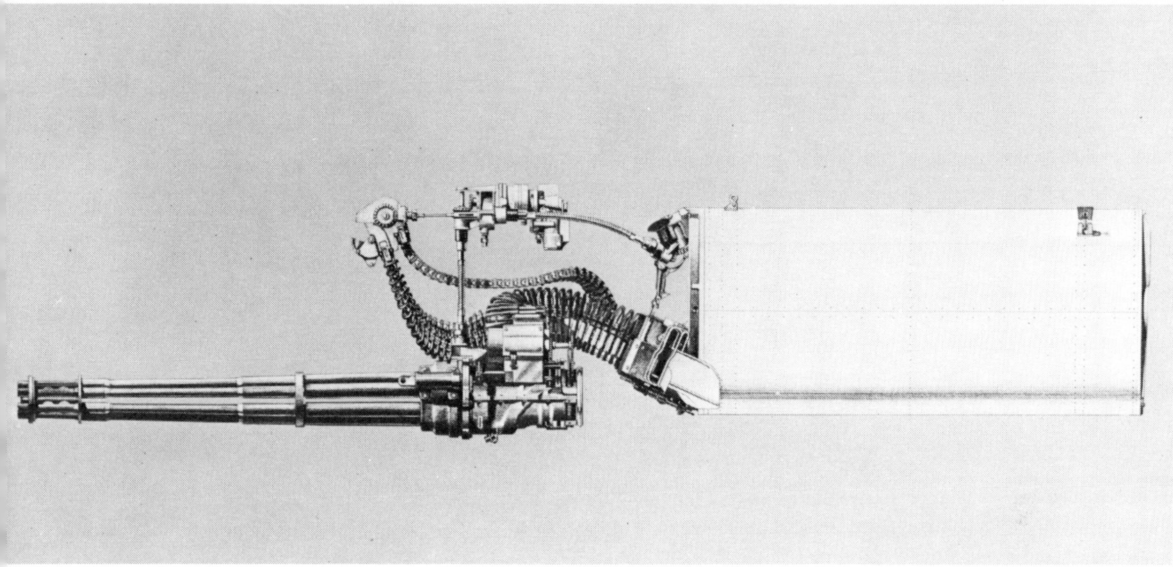

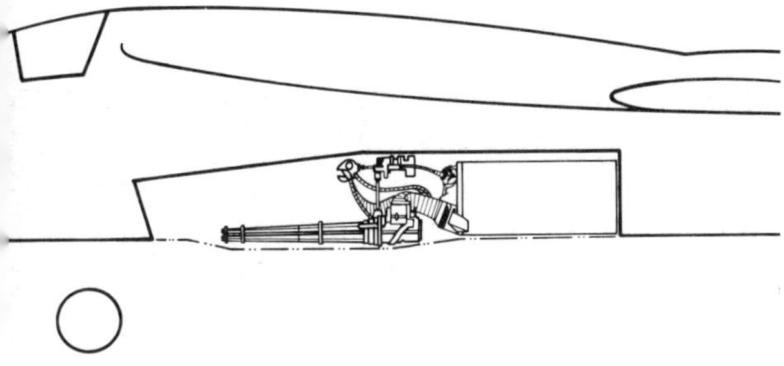

Top: This is the installation of the General Electric M61A1 20mm Vulcan gun in some models of F-111. The six-barrel gun weighs 252lb and fires 5,000 shots per minute. On the right is the drum housing an exceptional 2,084 rounds, weighing 1,039lb. Above is the hydraulic drive to the gun and linkless feed system, weighing 456lb. / *General Electric*

Above: This diagram shows how the M61A1 gun and its ammunition and feed system fit into the internal weapon bay. Aircraft fitted with the gun, mostly F-111A models, can be identified by the blister under the weapon bay. / *General Electric*

but before he could get this Voorhies had slumped violently to the ground, short of the runway. When the dust had cleared, the test pilot was seen to climb out of the damaged aircraft, take off his helmet and drop-kick it into the desert. Then, whilst being filmed by the cameras of all the major TV networks, he pounded the aircraft with his fists, yelling 'Dumbhead, dumbhead . . .' Later this extremely experienced pilot flew the F-111 spin programme.

One of GD's great band of flight-test engineers was Grover 'Ted' Tate. Compared with the dramatic years of the B-58 the F-111 was almost uneventful, but things did happen occasionally. His most memorable trip was in the first F-111B (pictured in this book). He recalls: 'John Norris and I ferried it from Edwards to Lakehurst Naval Air Station. From Edwards to Fort Worth we had perfect weather and a perfect airplane. Over Dallas our inertial nav system packed up, but the weather forecast for the rest of the journey was good so we pressed on. A few minutes later the oil pressure on No 1 engine quit indicating. As we crossed northern Louisiana Metro advised us that the weather was deteriorating, with a storm moving in across the coast. We did a lot of dead-reckoning, got ground-steers, took some advice from God and had a bit of luck thrown in, and eventually arrived over somewhere we thought might be the state of New Jersey. At this magic moment nothing in the aircraft was working except the clock.

'Fuel-warning lights were on, a vicious storm was all around us, heavy rain was slamming on the canopy and our radio was, to put it optimistically, intermittent. Finally we raised Lakehurst, advised them of our circumstances and begged for an immediate landing. Lakehurst advised us that their runway lights had been knocked out by the storm, the first 3,000 feet of the runway was closed and there was a lake of water on the rest. Other than that, we were welcome to be their guest. By this time the NO FUEL lights were brilliant. I favoured bailing out, but John didn't want to get wet in the heavy rain outside. We argued the point for a while, then headed inland and I went MAY DAY. We picked up a visual strobe on the horizon, got voice

Above: The flamboyantly painted first EF-111A, which did not have full operative systems. The EF-111A will have greater electronic capability than any other tactical aircraft./*Grumman*

Below: This former F-111A development aircraft is flying with completely different swing-wings of supercritical section (with flattish top and down-turned rear, as seen from the negative-incidence tip). NASA's Dryden Flight Research Center is testing the wing as part of its TACT (Transonic AirCraft Technology) programme, mainly with a view to civil-transport aircraft of the 1980s./*NASA*

TACT
TRANSONIC AIRCRAFT TECHNOLOGY
39778
USAF
NASA

contact with McGuire AFB, they just got a skin-paint on their radar, passed us a chandelle correction that exceeded our *g* limits and we were on finals — still with visibility outside absolutely nil. I identified our aircraft as an F-111B. McGuire tower asked me again the type of aircraft, and being rather busy with landing tasks snapped back "We're a damn big jet fighter". McGuire urbanely replied, "Roger, damn big jet fighter, you're cleared to land."

'We hydroplaned to a stop, totally out of gas and with nothing left that worked. John's clothes had become soaked, and he was so mad about the whole affair that he wore his dirty flight suit, helmet and all, to New York City and on the TWA flight home.'

Tate accompanied Grumman test pilot Don Evans on a mission that 'broke every airspace law known to man' to work a later B-model up to Mach 2 whilst loaded with Phoenix missiles; both engines flamed out at 45,000 feet, but they were eventually lit again at 7,000 and a landing made at Edwards 'somewhat ruffled'. Possibly the most dangerous F-111 trip he recalls was during drop-tank trials with an F-111A. Voorhies dropped the monster 600 US-gallon tank off its pylon at the appointed moment, only to have it return and angrily attack the aircraft. It hit the underside of the aft fuselage very hard, knocked out the right engine, ruptured the internal fuel cells and started a fire. Tate was ready to depart via the escape capsule, but Voorhies calmly landed back at Edwards. Then they were both subjected to all the psychiatric, blood and centrifuge tests that follow a major accident!

Having cut their teeth with a complex system on the B-58, GD were not daunted at the Air Force demand for an escape capsule. Pressurised and air-conditioned, it provides 'shirtsleeve' environmental conditions and safe escape even at zero altitude and zero airspeed, or at any point in the performance envelope. A press/squeeze/pull lever by the pilot's right knee triggers the system, which separates the module with explosive cutting cord, fires a rocket motor which thrusts it up over the tail and then releases the drogue and descent parachute. After a solf landing the module serves as a survival shelter on land or sea. On water the control column acts as a bilge-pump. To quote Grover Tate again:

'Col Henry Brown, a 17-victory P-51 ace, had been giving me pure hell for months about the "Micky Mouse, Chinese New Year escape system", which he was sure would fail to work at the crucial moment. Flying with Maj Joe Jordan, the F-104 altitude-record holder, he was firing the F-111A gun when gases in the weapon bay exploded. They triggered the capsule and rode down in comfort. On the next day I had to help the colonel write an article about his ejection for his Command magazine. It was full of superlatives of praise.'

It is surprising that Col Brown so distrusted the escape capsule, because he could not have failed to know Col Tom Wheeler, and Wheeler could practically sell anyone an F-111 for cash. I had the privilege of sharing a cockpit with him at Fort Worth — unfortunately, not in the air — where he was USAF Chief of F-111 Acceptance. He flew every F-111 delivered to the Air Force until late in the programme, and on 22 May 1967 he took off from Loring AFB, Maine, and flew to the Paris airshow without inflight refuelling and without external tanks. It was his greatest pleasure to demonstrate the TFR (terrain-following radar), the little pair of dishes under the nose that keep the F-111 skimming along the undulating ground, hopefully under the enemy radars, and even pull the aircraft smoothly over or around factory chimneys or TV masts. Called APQ-110 and a product of Texas Instruments, the F-111A TFR looks lost beneath the monster General Electric APQ-113 multi-mode radar that fills the nose and provides the navigator with all the pictorial information he needs for navigation and blind bombing. But the TFR can safely be entrusted with the crew's lives. This short extract from an essay by author Ernest K. Gann in *True* magazine, published in August 1971 after a trip with Wheeler (then a major), tells how:

'As you sink down from the heights, the mountains become ever more imposing and you remember a wager made the night before. If during the TFR run you manage to resist reaching for the stick and taking over from the automatic pilot, you will win two martinis . . . here are the 11,000-foot mountains approaching from directly ahead at a mere 553 miles per hour and you are looking *up* at them from the vantage of 200 feet. You indulge in some erratic swallowing and glance furtively at Wheeler, who is not looking out at all but thoughtfully studying an old aviation chart, vintage 1935 . . . As if cushioned softly against the face of the mountain, the One-Eleven rises with the slope . . . There is a saddle in the mountain ahead . . . trees higher than eye level, rocks, boulders, great chunks of very abrasive substance off both wing tips. Wheeler reaches for a knob. "Do you prefer your ride to be soft, medium or hard?" Your hand is curved like the talons of an eagle exactly one half inch ahead of the stick . . . The One-Eleven slips through the saddle and starts down the backside of the mountain, still at 200 feet and just under the speed of sound . . . Just as you are nearly resigned to continuing straight through the bottom of the valley on the direct route to Peking, the autopilot levels the nose and you slither across the rumpled surface like a giant manta ray seeking food.'

Above: One of the first production A-models with early base to fuselage, shown over desert near Nellis with wings at intermediate sweep in 1969. / *GD*

Having a ride with a pilot like Wheeler is great for building confidence in a weapon system that has suffered at the hands of the media. But if the hardware is genuinely suspect, truth will out. As the Navy said of the F-111B, 'If the pilots don't like it, it will never reach the fleet', and the same is true of any Western combat aircraft. So 17 July 1967 was a very important one for the One-Eleven. That was the day that Col Ivan H. Dethman collected an F-111A off the Fort Worth flight-line and flew it to Nellis AFB, near Las Vegas, where he commanded Detachment 1 of the 4481st Tactical Fighter Squadron. It was almost a year since the media had begun to assassinate the F-111 itself as well as the way it was bought, and the word 'lemon' had become a *sine qua non.* Dethman said he would stub out his cigar on whoever used the word in his earshot, but he had no illusions regarding the size of the job facing him. His crews were not going to have the advantage of an enthusiastic, or even unbiased, prior opinion of their new equipment. They were going to criticise the side-by-side seating. The fighter jocks were going to want more engine power, more manoeuvrability and a lower wing loading. So he took the positive line. He knew that the F-111A was a bomber, and the best tactical bomber in the world, by a very wide margin indeed. All he had to do was prove to the world that it did its job, and that it could generate enthusiasm among those who knew it best.

In a programme called 'Harvest Reaper' Dethman's outfit, despite having to entertain almost all the Top Brass in the Air Force, numerous politicians, newspapermen and hundreds of other visitors, put the first five aircraft to reach the inventory through the most intensive Intensive-Trials anyone could remember. The target was 30 hours flying by each aircraft in the month of September 1967, and though no two of the five were identical in systems, maintenance tasks or even flight envelope, the total achieved was over 304hr, or just over double the target. The following month the average per aircraft was 59.7. Even at this time the One-Elevens were still coming off the line with small differences, but their

Head-on view of an early F-111A in the folded configuration./*GD*

main parameters were firm. They were big, painted in green and ochre camouflage, with black undersides, always seemed ready to go, and landed back on the ground with a gentle cushioning feeling without the pilot having to flare to arrest the sink. Crew-chiefs and ground crews liked the big bird so much that the Nellis re-enlistment rate went sharply upwards.

Altogether GD built 141 A-models, including the first 30 which were all development aircraft, all different, and never brought up to operational standard. Though powered by the original P-3 engine, the A-model can carry 32,500lb of internal fuel, two nuclear weapons or two 750lb conventional bombs in its internal bay and a wide assortment of external stores on eight pylons on the swing wings. If the F-111 were being designed today more of the pylons might be on the fuselage. As it is, the four inboard pylons pivot as the wing sweeps to remain aligned with the local airstream, while the four outer pylons are jettisonable, non-swivelling and not often used. A typical load of 'iron bombs' is two triplets of 500-pounders on each of the four inboard pylons, which at a true unit weight of about 580lb aggregates to 13,920lb. Heavier loads can be carried, pushing the takeoff weight well beyond the admitted 92,500lb, but with just the load given above the takeoff run is around 4,000 feet on concrete, maximum Lo speed about 490 knots (564 mph), ceiling fractionally over 30,000 feet, and Hi-Lo-Hi radius about 600 miles.

These figures are down on SOR-183 but stand up pretty well in comparison with anything else. Later models of One-Eleven are more powerful and have higher performance. Admittedly, this meant that the Fort Worth line never really got into its mass-output stride, as had been hoped, and instead had to build small runs of high-priced aircraft of at least eight different sub-types. On the other hand, the aircraft got better and better. Even the original A-model can still do tactical attack missions that are beyond the capability of any previous aircraft in the West. In south-east Asia it proved it.

Below: Later F-111A (66-0018) with intermediate base to fuselage but full operational equipment, pictured in maximum-sweep configuration. / *GD*

Proof of the Pudding

On 6 August 1945 a US Army Air Force B-29 dropped a primitive fission bomb on the Japanese city of Hiroshima. It shattered not only the city but most prior theories regarding aerial attack. Henceforth it would be possible for single aircraft to mount damaging raids on large and important targets, which previously could not be hit sufficiently hard except by armadas of hundreds of bombers. But this presupposed the use of nuclear weapons. As the 'massive retaliation' policy faded, to be replaced by 'flexible response' and similar strategies involving conventional weapons, the notion of single-plane attacks likewise played a smaller and smaller part in the overall scenarios for possible future wars. Then the F-111A entered service, and it brought with it a dramatic new capability. This one type of aircraft for the first time made the one-plane attack with conventional weapons a practical possibility, which has again transformed the planning of airstaffs. The technology of the F-111 combines such reliability — expressed in what the air forces call a 'low abort rate' — and such pinpoint delivery accuracy that major attacks can be planned with single aircraft. It seemed that this would be possible at the paper stage, and seemed even more likely in the early days of flight test and evaluation at Edwards and Eglin, and in mission training over the desert and mountain bombing ranges near Nellis. But it took the test of combat in the SEA theatre (South-East Asia) to prove it.

The story goes that, on arriving at Takhli AB in northern Thailand, an F-111 pilot was asked 'Do you people have smart bombs?' His reply was 'No, but we've got smart airplanes.' Today even the media know this.

In proving it the crews of the F-111s also proved, more completely than has ever been proved anywhere else in the world, that it is possible for tough, seasoned aircrew, confident in their own skill and experience, to sit back and let boxes of electronics take over absolutely and completely in circumstances where the slightest technical fault would mean instant death. Indeed it is doubtful that humans have ever before so completely trusted their lives to what could popularly be described as automation, in this case a mix of a radar and a computer. Maybe it takes exceptional men to do this, and the reader can judge from this chapter the kind of ice-cool nerve needed; but today there are many F-111 crews in many places for whom this kind of utter trust in a man-made system has become an accepted part of a regular job. Of course, the submariner trusts that his hull will not be crushed, and the ordinary pilot trusts that his wing will not come off. Operational flying in the F-111 is something very, very different!

When the F-111 was being developed, the United States was becoming embroiled in a military campaign in SEA to help support the government of South Vietnam. At first this was little more than a limited or 'brushfire' war. The United States had developed specifications for aircraft to fight in such wars, and the first to be built was the OV-10 Bronco, a turboprop STOL machine capable of flying light missions against a primitive enemy and of evacuating casualties from a ground battle. The F-111 was a quite different animal, designed for the top league of warfare between sophisticated enemies. It did not at first seem appropriate to the escalating warfare in Vietnam, but during 1968 the decision was taken in the Pentagon — probably purely at the political level — to send a token force of the controversial swing-wing machines to see how they proved out. Those who took the decision may have been friends or foes of the F-111, and even at that time it had plenty of both. In fact the decision was welcomed by the One-Eleven's friends, because it at last offered a chance for this most capable aircraft to prove what it could do.

In early 1968 the F-111 was still a very new and not fully developed weapon system. Though the terribly difficult engineering development and flight-test programme on the initial F-111A model was mainly complete, Air Force Cat I testing was still in progress and Cat II and III had not even started. It was thus nothing like ready for combat duty. On the other hand those close to the aircraft were so keen to show what it could do that there was not much contrary argument. In fact the pilot of the first aircraft to take off from Nellis for Thailand was Col Dethman, now CO of the 428th TFS. From the start he agreed that the aircraft

Right: Skiing up the face of an isolated peak. The crew can judge the shape of obstructions ahead, and can elect to go round rather than over. With the ride set to 'Hard' this peak would pull the aircraft over the top with powerful negative *g*. / *USAF*

should always operate the way it had been intended to, with no external assistance. So Dethman flew across the Pacific on the internal inertial system, and internal fuel. This alone is a task that no previous tactical aircraft could accomplish, and on sectors such as the 3,000 miles from Hawaii to Guam the consequence of bad navigation or inadequate fuel is self-evident.

The plan was to deploy six aircraft in a 'combat evaluation' called Combat Lancer. Though six aircraft at an advanced base half a world away from the United States is not a cost/effective operation, it was expected that much would be learned. Plenty was, but by sheer bad luck — which really dogged the One Eleven's early years — the balance sheet looked a disaster. Dethman landed at Takhli AB on 17 March 1968, and missions began a few days later after a few training sorties in the SEA environment. On 28 March an F-111 went out on a combat mission and failed to return. On 30 March another did likewise, and a third vanished on 27 April.

To lose half the force in a month against what were at that time primitive defences was discouraging. It was not possible to get even a single clue about the first two losses, because the crews maintained radio silence and the wrecks were never found. The only indication was that the aircraft appeared not to have been shot down in the target area, and this belief was strengthened by the report of the third crew, who managed to eject. Their evidence, and examination of the wreckage, pointed strongly to gross failure of the left tailplane (horizontal stabiliser) control system. First inspection revealed a large lump of tank sealant on the tailplane power unit, but more thorough examination brought to light the true cause. Inside the unit the main control-valve rod had failed in fatigue, snapping in two at a welded joint. This had immediately driven the left tailplane to the limit of its negative travel, causing sudden violent roll and pitch-up, whilst slamming the stick back into the pilot's stomach. On 8 May, before the true cause had been established in Thailand, a Nellis F-111A went violently out of control, the crew surviving. This was quickly and positively traced to failure of the same weld.

By this time the Air Force had lost six One-Elevens in just over 5,000 hours. This put it ahead of all other combat types in flight safety; but when it is recognised that at least two, probably four and possibly five of the six write-offs, had been caused by failure of a weld, which could have been prevented at the start by more effective quality control, the good safety record ought to have been many times better.

Unfortunately, this was not the end of the structural problems. Though the weld problem had been discovered and rectified, with new valve rods throughout the F-111A fleet, the three survivors in Thailand were withdrawn after a mere 55 combat missions. Not very much had been learned beyond the fact that serviceability was outstandingly good and the North Vietnamese radars hardly ever 'saw' an F-111A for an instant. Not one of the Combat Lancer aircraft had as much as one bullet hole, and there was no evidence that any of the lost aircraft had been hit by any defences either. Yet, on balance, the combat deployment had not been an unqualified success, and the return of three aircraft out of six with not very much outwardly to show for it was not the 'outstanding combat capability' the Air Force had wished to demonstrate. Then GD suddenly found fatigue cracks in the most critical place: the swing-

Above left: Late-production F-111A about to receive fuel from a KC-135 of an Air Refueling Squadron. Thanks to its immense internal capacity the F-111 hardly ever needs to refuel in flight. / *GD*

Above: An F-111A of the 474th TFW, with four pylons equipped for normal loads up to 12,000lb of conventional bombs. / *USAF*

wing pivots. By sheer good luck they were discovered on the ground, during fatigue testing.

GD's F-111 design had earned good marks because, unlike its rival, its wing box was made from steel instead of the supposedly less proven titanium. The material chosen was Ladish D6AC steel, a long-established and conservative high-strength steel used in many critical aircraft parts and also for solid-propellant rocket cases. It had a good history, and GD had ample experience of how to design and test primary structure. Most parts of the wing box are welded together but the specially heat-treated top plate is joined by bolts. This demands rigorous quality-control, with every hole precisely drilled, reamed and polished to avoid any sudden discontinuity or scratch that might induce fatigue. The painstaking care lavished on the F-111A wing box can be imagined, yet in late 1968 GD discovered a series of fine fatigue cracks radiating from one of the bolt-holes on the fatigue-test airframe. This was extremely serious, because anything involving fatigue is subject to statistical scatter which means that it might rear its head much sooner in a different airframe. All One-Elevens were temporarily restricted to 3.5 *g* until their swing-wings had been removed, the box opened up and the lower corners reinforced by three 8in gussets, which added 500lb to the weight and cost several million dollars.

Was this the last of the problems? Far from it! On 22 December 1969 an F-111A pulling out from a low-level attack shed its complete left swing-wing. The crew did not escape. Inspection showed that the speed, load-factor and aircraft weight were all quite normal, and that the wing box had been modified. It also showed that the failure had not been in the wing box but in the wing pivot fitting, the part on the inner end of the swing-wing pivoted to the box by a precision-fitted tubular pin about as large as a gallon of paint. Amazingly, the lower plate of this wing fitting was found to have contained a large flaw, a crater about an inch long and almost the full depth of the plate. It had escaped being detected by magnetic testing, ultrasonic testing and three visual inspections.

So important was the F-111 judged that, after this troubled structural history (in the very material previously thought to be taken for granted as troublefree), the unprecedented decision was taken to undertake detailed non-destructive inspection and proof testing of every aircraft. At a cost of millions of dollars, four test facilities were constructed for this purpose, two at Fort Worth, one at Waco and the fourth at Sacramento. Here every F-111 was subjected to a proof test at the 56° wing position at a refrigerated temperature of -40°C (-40°F), once at the design positive load of 7.33 *g* and once at the negative load of -2.4 *g*. No further trouble was met with the wing pivots, but two potential crashes happened on the ground, in one case when the left tailplane pivot shaft fractured and in the other when the whole lower plate of the wing box split across at only just over half the

Above: USAF copy-print of a 428 TFS aircraft letting go CBU-58 cluster bomb dispensers; 16 is a normal load. / *USAF*

Right: Some of the largest steel parts in the F-111A. Several, including the wing carry-through box and wing pivot fitting, gave prolonged trouble in 1966-70. / *GD*

Far right: Close-up of left inlet duct and pivot of wing carry-through box in an uncompleted F-111A. / *GD*

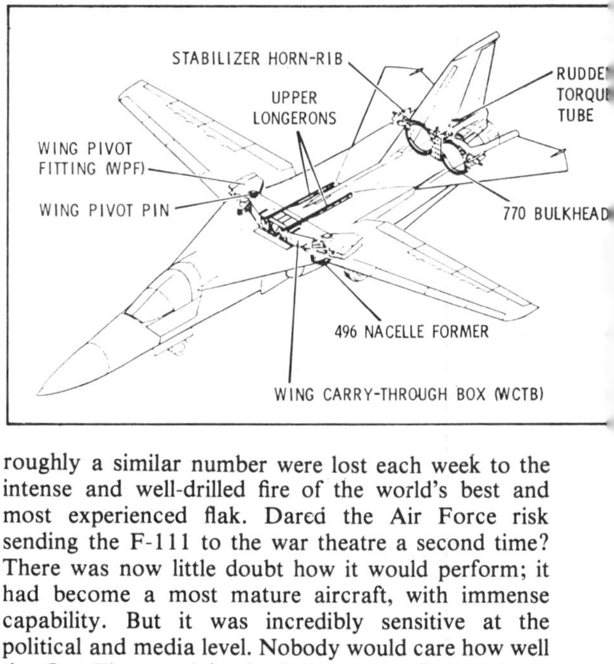

7.33 *g* load. No such structural test programme had ever been carried out previously, and the F-111 remains unique in this respect.

During the early 1970s the air war in SEA changed dramatically in scope and character. From being a relatively limited conflict, it escalated until the defences of North Vietnam against air attack were among the greatest ever assembled. In terms of density of 'triple A' (Anti-Aircraft Artillery) firepower it exceeded that of the Ruhr in World War 2, while the SAM (Surface-to-Air Missile) defences were even more unprecedented, with a dense net of launch sites, each with eight or more launchers, and a vast stockpile and supply system for missiles. Radar coverage was likewise fantastic. One F-111 pilot said 'There is no such thing as coming in "under the radar" in the Red River delta. The place is so small, so heavily defended, so flat, that they are looking in all directions for attacks all the time.' Though the Soviet-supplied V-750 series SAM system (called SA-2 *Guideline* by the West) was primitive and not very deadly to a well-equipped tactical aircraft, the sheer volume of SAM fire caused heavy casualties to American aircraft, and roughly a similar number were lost each week to the intense and well-drilled fire of the world's best and most experienced flak. Dared the Air Force risk sending the F-111 to the war theatre a second time? There was now little doubt how it would perform; it had become a most mature aircraft, with immense capability. But it was incredibly sensitive at the political and media level. Nobody would care how well the One-Eleven might do, but an aircraft shot down would make headlines from coast to coast.

Eventually the Pentagon decided to grasp this particular nettle, and it detached the other two squadrons of the Nellis-based 474th TFW to Takhli in the autumn (fall) of 1972. The units were the 429th and 430th TFS, and they flew across in groups in September 1972, fully operational from the start. In fact the very first pair of One-Elevens to arrive on this

detachment were off again on a mission within three hours of their arrival from the United States; and only one came back. At first it looked as if this was the same old story of rotten bad luck, and to this day nobody knows what happened to that missing aircraft. But in fact the SEA deployment of the 429th and 430th was — at last — a story of tremendous accomplishment, ranking with the greatest bombing missions of all time. Because it was a story of positive achievement by the One-Eleven it got scant coverage by the media. Few people, even in the United States, know what those 48 aircraft did in the final five months of the Vietnam war. Whatever one may feel about the rights or wrongs of the US involvement, no combat aircraft in history have ever flown tougher, more courageous or more successful missions.

Almost throughout, the operational philosophy was the Lo penetration using iron or cluster bombs. The F-111 could have bombed from high level or attacked at various heights using smart bombs or missiles, but more than 98 per cent of the 3,500-odd missions were flown at Lo level, on the TFR, carrying no stores but 500, 750 or 2,000lb iron bombs or cluster bombs. The usual number of stores loaded was 12 to 16. With some aircraft this might not have been very effective, but with the F-111 there was the very important difference that the load fell directly on the target. Analysts of the 7th Air Force rated overall F-111A bombing accuracy as extremely close to that achieved by other aircraft using smart bombs. The difference lay in the fact that, while the smart bombs were steered all the way to their targets in fine weather, the One-Elevens went out at night or in foul weather. Only one mission was cancelled because of weather, and that was at the height of the worst monsoon rain, when flying was virtually impossible.

Soon the well-worn F-111As had established a reputation for serviceability, low abort rate and mission effectiveness in a class of their own. The only other type of aircraft that could approach them in their all-weather navigation and attack capability was the Navy A-6 Intruder, and this was not the equal of the One-Eleven in circular error, over-target speed or ability to operate without assistance. When working out the cost/effectiveness of different types of aircraft the costs of the F-111A were total costs, while those for other types required the addition of costs and logistic support for KC-135 tankers; EF-105 Wild Weasel, EB-66 Destroyer or EA-6 electronic support aircraft; F-4 fighters flying MiG-CAP top cover; other F-4s seeding the sky ahead with 'chaff'; a varied fleet of communications and relay aircraft; various air traffic or forward air control aircraft; and a heavy commitment of fixed- and rotary-wing rescue aircraft. In contrast, the F-111A worked alone, but hit the hardest targets.

According to the CO of the 474th TFW, Col William R. Nelson: 'We were able to take out all the point targets in single missions by single airplanes . . . There was one target that was particularly important that was assigned strictly to F-111s. We were the only aircraft which struck this target. We flew against it every night, several times a night, for about two weeks. It was surrounded by city. Every one of our bombs was in the target area. The destruction was heavy, and there was no collateral damage . . . We were instructed that if we were not positive that we were going to put our bombs on the target we would not drop them. I actually had some people go all the way into the Hanoi area and come home with their bombs. But on this target — it was a highly populated area — the bombs were all dropped at Lo level, they were all dropped at high speed, they were all dropped using our radar and computer system, and they were all dropped on target.'

What was it like, to fly the One-Eleven 'for real'? No other aircraft has ever actually gone to war with the technology of the 1960s. MiG-21s, Phantoms, Mirages — these are the technology of the 1950s, and these aircraft cannot do what the One-Eleven can do. In one sentence, the new capability is that the technology of the 1960s allows a blind first-pass attack. This means that the attacking aircraft can crank into its computer the geographical position of its target, fly there (making any desired dog-legs or evasive action en route) and come straight over the target at close to the speed of sound and hit the target with ordinary iron bombs. It makes no difference whether it is day or night, or clear visibility or solid 'clag'. The pilot or navigator merely sit in their seats and monitor what goes on; they do not even have to see the target. The blind first-pass attack is still beyond the capability of most aircraft in the world's air forces, two of the exceptions being the RAF version of the Jaguar (which does not have a TFR) and the Tornado.

Left: Col Ivan Dethman says a few final words at the briefing for Detachment 1 of 428 TFS before departing from Nellis for Thailand in March 1968. / *USAF Official*

Below left: Somewhere over the Pacific, Detachment 1 of 428 TFS is seen heading west for SE Asia in March 1968. All six aircraft can be seen, the other three being in the far distance clustered round a KC-135. / *USAF Official*

Below: Police guard an F-111A of Detachment 1 on arrival at Takhli AB on 17 March 1968. This was the first 'Aardvark' to be seen outside the United States. / *USAF Official*

A few other types are also members of this exclusive club, one probably being the Soviet Su-19 *Fencer,* but the F-111A remains the only aircraft so far that has done it over a long period in actual warfare, where the hostile airspace is genuine.

Making a blind first-pass attack in good weather, with the pilot looking ahead through a head-up display (HUD) and exerting a direct control over the proceedings, is what might be called sporty. I have done this in a Jaguar, and would love the chance to do it again. But doing it in a One-Eleven in an intensely hostile airspace, in foul weather, is something utterly different. The crews in SEA called their ground-hugging profile 'skiing'. Some people who are not regular crew-members have lost their nerve in the right seat of a One-Eleven on a skiing run— and need not feel ashamed of the fact. Gann (p.46) did not, but got his feelings across. But not one of those given a demonstration TFR ride ever experienced anything like the missions that were flown in Vietnam. And the reason the missions were so tough is that there were surveillance radars on every hill and eminence, and Triple-A gunlaying radars and SAM radars were dotted throughout the flatlands. On every mission the strident passive warning systems in the One-Eleven cockpit would light up as soon as the aircraft crossed into Communist territory. No matter how closely the 'skier' hugged the ground, it would be rare indeed for all the radar warnings to be extinguished. Of course, it takes pretty good radars to lock-on to a single sharp-nosed swing-wing skier speeding along by himself. The F-111A countermeasures in 1972 merely indicated the receipt of hostile radar energy (and doubtless gave the frequency, waveform and pulse-repetition frequency to assist the on-board jammers). It did not tell the crew what the enemy could actually see, and to what extent they could aim their Triple-A and guide their missiles.

Above: Ready for the first combat mission, an aircraft of Detachment 1 waits in its revetment at Takhli AB with twelve 500lb bombs loaded. / *USAF Official*

Above right: Preflight inspection prior to a mission on 15 September 1968. There was an armed guard at every occupied revetment. Note the ALQ-87 active ECM pod immediately behind the man in the white shirt. / *USAF Official*

It added up to an environment of frightening hostility, posing a very great challenge even to a One-Eleven. Six aircraft were lost, four from the 430th TFS and two from the 429th, all in highly defended areas. It is amazing that losses were not higher. The defences — which crews called 'unbelievable' — knew the aircraft were coming and knew what the targets would be. The only thing on the attacker's side was that each crew could plan its own mission, try to make it different and unexpected, and was operating alone. Each crew might spend six hours one night planning the mission it would fly the next night. It would be a matter of making the flight plan as random as possible, but with each element of randomness precisely judged to try to throw the defences off-key. Only rarely did senior staff officers or planners offer words of advice, to suggest how to correct what appeared to them to be an error. The crew did not have to take the advice; it was their mission.

Before they left Nellis the crews were experienced, and wholly confident of their electronics. But skiing at Mach 0.9 at 200 feet is obviously inherently dangerous, even over the deserts of the American south-west. To ease the psychological stress the TFR has two channels, and, if either malfunctions or quits, a warning shows in the cockpit. The flight-control system is triply redundant; if one channel quits a

warning shows, and if it malfunctions and injects a faulty command it is 'outvoted' by the remaining two. On top of this, any malfunction in TFR flight triggers an automatic 2.4 or 3 *g* pull-up to a safe altitude. Yet, despite these safeguards, the knowledge that the ground is 0.25 second away is hard to forget. In the Takhli Officers' Club there was a notice which reminded crews that the effectiveness of SAM shots averaged less than 15 per cent, and that of Triple-A was less than five per cent, whereas the effectiveness of the ground was always 100 per cent. Nobody was likely to forget it. A Weapon Systems Officer, Lt Steve Glass, admitted 'We were always nervous, no matter where we were targeted'.

In 1973 the US *Air Force Magazine* sent Wayne Thomis to Thailand to talk to the crews. Capt Jackie Crouch, a former F-105 pilot on two tours of SEA operations, explained the intensity of the nervous problem: 'Think about flying around in daylight and good weather only 200 feet above the ground, and going up and down over hills and into valleys, keeping this height. Now do this at night, in mountains and in heavy cloud when you can't see anything outside the cockpit. This is really, really exciting, even without the enemy threat.

'It takes real discipline to come up over these mountains, as we did at night, out on top of the cloud layer in the moonlight. We'd see those jagged peaks all round us poking through the cloud tops, and we'd have to put the nose down back into that mist. And as we went down the moonlight would fade, and the cloud get darker, and we'd know we were descending far below those peaks and were depending on our radars and our autopilots — and with Hanoi coming up. I won't say that I wasn't worried.

'One night, when the weather was *very* bad, I was in cloud for the last eleven minutes before bombs away. That means at the lowest levels of the whole flight, at 250 or 200 feet going up and down the hills. We didn't see a thing outside the cockpit, not even after the bombs left us. For me, this thing was really remarkable. Even now I can't explain how fantastic it was . . . the confidence I gained in the airplane, it made a believer out of me. Given a choice on a night strike of going in Hi or going in Lo, I'll take Lo every time. And I'll go anywhere in the F-111'.

A left-seater with the 429th, Major Carlos Higgins, said the enemy respected the One-Eleven:
'They call it "Whispering Death". This indicates the surprise with which we hit them. When we bombed in bad weather or rain, we wondered whether they could hear us in advance at all. They must have been surprised when, without warning, the bombs flashed on impact in downpours, as they often did. We thought the bomb flashes were often their first clue — other than their radar — of where we were.

'Once you were Lo, you had to monitor everything. You had to be thinking, and looking — by radar — as far ahead as possible so you would know what the terrain was like, and you had to count on this information from your right-seat man whose radar is better than the small vertical indicator scope on the pilot side. You counted on your right-seat man to keep giving you word of obstructions seen on his scope, so you could be sure the autopilot was obeying the TFR.'

There could hardly be a more full-time job than making a TFR attack, so the decree of 7th Air Force HQ in November 1972 that crews must radio their call-sign and position at course-change points was not welcome (an automatic digital broadcast of selected mission and aircraft data would have been more appropriate). Crews were so busy they often just let it

go. According to Higgins, 'If you missed making a position report by five minutes, they'd call you. Often I'd come back with "Jumbo Two Four, still alive" and let it go at that.'

In the final stages of the war the bombing was forbidden north of 20°N latitude while peace talks were sought. During this period vast quantities of SAMs and AAA ammunition were brought in. But when bombing was resumed on 18 December 1972 the weather was so bad the North Vietnamese thought no attack was possible. Lt Glass recalled:

'The delta weather was way down, ceiling 200 feet or thereabouts and cloud piled up to 28,000 feet . . . We came skiing down the mountains and plunged out into the open under the lower edge of the overcast, and it seemed to us the entire Hanoi valley was lit up like Las Vegas. Hanoi was bright with neon and street lights and the port of Haiphong was aglow in the distance. On the roads leading out of town and on the mountain switchbacks to the south the truck headlights were blazing like strings of pearls.

'We happened to arrive about ten minutes to eight in the evening, Hanoi time. We were coming so fast, we were almost on release point before any of those lights started going out. Sections of the town blacked out one at a time, and we knew sirens were screaming and somebody down there was pulling master switches, even as the bombs left us.' (Bombing Hanoi, of course, was forbidden; bombs had to go on specific and proven military targets, and as noted earlier nobody was allowed to release if there was the slightest element of doubt.)

Top left: This aircraft of Detachment 1 is leaving its revetment at Takhli with only the outer pylons loaded and with ALQ-87 under the fuselage. The F-111A did not need the electronic help of the EB-66 seen in the background. / *USAF Official*

Above: An aircraft of Detachment 1 returns to Takhli Royal Thai AFB after a combat mission. The gun is fitted, as shown by the bulge behind the nose gear; it is seen clearly in the lower picture on p.56. / *USAF Official*

Top right: Two of the survivors of the second SE Asia tour undergoing hangar maintenance at Takhli immediately before the last combat mission on 15 August 1973. Note the repainted tails with code letters HS. / *USAF Official*

Such attacks caught the defences napping, but the B-52s were not so lucky. On the three days 18-20 December 1972 Air Force ECM crews watching high overhead counted 600 SAM firings. Of these about 120 were at night, nearly all at the B-52 altitude of 28,000-35,000 feet. Some even came after the lone skiers. Glass said 'On those nights we saw plenty of fireballs. Once I looked back after "bombs away" and saw four rising together, a salvo. And over the common radio frequency, which we all monitored, the countermeasures watch never stopped talking, reporting SAM tracks. I heard him call out at least 50 before were were out of range.'

So serious were the losses and damage to B-52s that, after three days, 7th Air Force turned the One-Elevens loose against the SAM sites. Each site was shaped like a star, with radar and control trailers at the centre and launchers at the points — in bad weather or at night an almost impossible target for ordinary aircraft, but routine for the One-Elevens. Captain Paul Sperry, right-seater for Major Higgins, reported 'We got counts of 80 to 120 firings during each of the early nights. After we got to work they dropped to 28 on 21 December and 18 on 22 December.' And Crouch added 'There were nights after that when not a single SAM was fired. I think we convinced 7th Air Force of our accuracy and our value after those four tough days.' Not many aircraft can go out to hit SAM sites.

Without exception, the crews were surprised at their ability to go in through a sky full of flak and missiles, hit the target, and come out without a scratch. Time after time Triple-A shells would pass so close that the crew would hear the 'whack' of the shockwaves. 'One night', Glass related, 'five rounds went by so close we felt their passage and heard it. We thought we'd been hit. But we looked over the panel and nothing flickered. When we got home, inspection turned up nothing.' Crews averaged 45 to 53 missions, and by actual count there were fewer than ten hits taken by the total F-111A force (except the six aircraft missing). Only once did an aircraft make a precautionary landing away from Takhli. Putting down at Udorn, the aircraft was inspected, the routes of the SAM splinters traced, the aircraft certified airworthy and the entry holes taped over.

In the final reckoning, the F-111A force flew more than 4,000 sorties between October 1972 and March 1973. The loss rate was 0.15 per cent, the only known cause being SAM hits (four remain completely unknown, but were almost certainly due to enemy fire). No other aircraft in history has proved so survivable. Neither is there any record of any other aircraft sustaining prolonged combat missions with an abort rate as low as 0.85 per cent. Many aircraft completed 100 combat missions in less than six months, without any damage or aborted mission and with only routine servicing. (In July 1973 a crew of the 429th, Captain Christopher F. Russo, aircraft commander, and 1st Lt Charles R. Foster, WSO, likewise completed 100 F-111 combat missions, 52 against N. Vietnam and 48 against points in Laos.) Altogether it added up to the significant fact that black boxes have revolutionised conventional bombing, provided that they are managed by men of skill and courage.

'Aardvark' in Aussie

Perhaps because of its troubled early life, the F-111 never received a name. Generally it is called the 'One-Eleven', like the British civil airliner, though a few even speak of the 'One-one-one' (bringing memories of Heinkels of long ago). The only popular name is an unofficial one; somebody thought that the long, probing nose resembled the African ant-eater called an aardvark, maybe so that in any alphabetical list the F-111 would be well placed. Of course, the aircraft is found only in the United States, England, Australia and South Korea, where there are no aardvarks outside zoos. I consider this extremely important aircraft deserved an official name — such as Attacker, Penetrator or Striker — and definitely a new number in the 1962 combined-services numbering system, to lift it right out of the controversial past. Obviously, this number should have been prefaced by an A or B, rather than the inappropriate F.

Never was an aircraft more inadequately designated, but this was the least of the problems for the only export customer. In 1963 GD told me they expected 'between eight and 12' countries to buy different versions of the F-111, and some of the long-term company planning was based on these foreign sales adding to the domestic programme of 1,726 aircraft to take the total to considerably more than 2,000. On a more nebulous basis it was planned that the existing aircraft should during the 1970s lead to a directly derived further generation to continue production in the 1980s.

With the backing of the US armed forces, which at that time had shown no inclination to buy jet-lift V/STOL aircraft, GD expected to sell the F-111 widely in Europe no matter whether the numerous V/STOL projects of the early 1960s came to fruition or not. There were several other prime sales targets, and none was more important than Australia, which wanted to replace its ageing Canberras. In 1958-62 it had seemed self-evident in the Department of Defence in Canberra that the RAAF's Canberra-replacement would be the same as that of the RAF: the purpose-designed TSR.2. Several fruitful discussions were held at air force and ministerial level, and there is no reason to doubt that the British low-level reconnaissance bomber would have met all the needs of the RAAF. But by 1962 the Australians were becoming very interested in the American TFX programme, which appeared to offer something extra. It used this at first as a stick with which to beat the British and thus minimise TSR.2 development delays or escalation in costs. In February 1963 the Aussie Minister for Air, David Fairbairn, said 'The RAAF cannot wait for TSR.2. It must replace its Canberras within two years.' He did not mention the exciting TFX but said his department was looking at aircraft that could be bought direct from current production, and named the A-5 Vigilante and F-4 Phantom from the United States and Mirage IVA from France.

In August Britain announced increases in the TSR.2 order-book, and the following month British Air Minister Hugh Fraser went to Canberra to offer 25 aircraft at a price, without spares, of a mere £65 million, with deliveries alternating with those to the RAF, plus an option of a V-bomber loan in the interim. But by this time the British Labour Party had singled out TSR.2 as a programme they wished to cancel, for reasons that were emotional rather than rational. The Australians recognised that they dared not plan their defence strategy for the next 30 years on a programme that might suddenly be terminated, and immediately sent a team to Washington to talk F-111. On 24 October Fairbairn very quickly signed for 24 of the American aircraft at a price of US$90,749,040, for delivery between July and November 1968. Obviously the over-riding need to replace the Canberras 'within two years' had been forgotten.

Prime Minister Menzies heaved a sigh of relief. The F-111 was American, with massive procurement planned for both the US Air Force and Navy, so it was taken for granted it could be relied upon to be ready on the planned day and at the right price. He said 'No government could spend money on anything else'; and his Defence Minister Athol Townley, stung by someone who said the price might escalate, retorted 'There is far more chance it will be reduced, because our figure is based on the present production run'. Hundreds more foreign sales were confidently predicted. Nobody bothered to write a single clause in the contract to offer protection in the event of trouble.

Needled at the apparent great success of the F-111 purchase by the ruling Liberal-Country Party the opposition Labor Party immediately denounced it as

Right: Unusual formation, for display purposes, by four Amberley Aardvarks. Each aircraft is carrying two practice-bomb containers. / *RAAF*

Above: Head-on we get a good view of the bird's fat tyres, which last on average more than 100 missions, and unusual width of fuselage for so fast an aicraft. Taxiing lamps appear to be mandatory even in bright sunshine. / *RAAF*

Left: They call this bit of Queensland, near Brisbane, Australia's Gold Coast (for the best of reasons). Nos 1 and 6 Sqn are based a few minutes away as the F-111C flies. / *RAAF*

'a vote-catching bid' (which gradually came to seem extremely ironic). To help tide over the former Canberra squadrons in 1963-68 the US government offered the lease of B-47E bombers withdrawn from service with Strategic Air Command, but the RAAF realised it would incur large costs on top of the hire charges in trying to put this aircraft into operational service. In July 1966 new Minister for Air Peter Howson announced that in early 1970 six of the RAAF swing-wingers would be flown back to the USA for conversion as RF-111 reconnaissance aircraft. Conversion would involve removing the weapon-bay doors and fitting into the weapon bay a large multi-sensor pod for surveillance to several miles on each side of the flight path with cameras, radar and infra-red linescan (which records TV-type pictures in which cooler regions are darker shades and hotter places look light). Howson said 'The striking power of the RAAF will not be affected, because the six modified aircraft can be reconverted to the strike role within hours'. The RF-111A, as the proposed US Air Force reconnaissance aircraft was designated, had been announced in December 1965. With exceptional design range and good penetrative capability the F-111 seemed well suited for use as an advanced reconnaissance machine, and a major development programme was put in hand for the multi-sensor pallet, with bulged radome and optical/IR viewing windows, and for a digital computer to manage the systems and relieve the workload on the weapon-system operator. The 11th F-111A was rebuilt as the prototype RF-111A, flying on 17 December 1967. Altogether the programme cost US$118m, and the RAAF paid additional sums in its regular progress payments towards their expected RF-111 costs of $A7,980,000. But the prototype was the only RF model ever to fly. Neither the USAF nor the RAAF has yet received a single reconnaissance pallet, though as described later the RAAF still wishes to.

In December 1966 the designation of the Australian aircraft was announced, logically enough, as F-111C. It was disclosed that it would have the long-span high-lift wing of the Navy F-111B and a new landing gear with stronger legs and larger tyres and brakes, such as was then being developed for two further models, the FB-111 and F-111K. These features were claimed to 'significantly increase the range and payload' compared with the F-111A. Another difference is that, unlike USAF tactical versions, the stick in the right cockpit can be unclipped and removed. In February 1968 Peter Howson announced that F-111C training equipment had been flown in to RAAF Amberley, Queensland, where US-trained NCO instructors were already teaching future ground crews. Several large systems rigs that duplicated aircraft functioning systems were in full use, a flight simulator was on order, and after further training courses in the USA had been completed the whole ground staff would be ready to receive the 24 aircraft in September-December (a slippage in timing of about two months). But in May 1968 doubt had been cast on even this schedule by the losses of aircraft in the USA and Vietnam, and a specialist team led by Air Vice-Marshal E. Hey, Air Member for Technical Services, spent three weeks talking in Washington, at GD and at Nellis. The Minister for Air, Gordon Freeth, said that Australia would not accept delivery until the F-111Cs had been 'thoroughly inspected and tested'. Allen Fairhall, Minister for Defence, emphasised his country's problems: 'While it may be competent for the United States authorities to put their aircraft into service within a few months on a reduced period of test, they have readily available to them facilities for structural modification, whereas if a need should arise for modification to our aircraft we would face tremendous difficulties'. What made this ironic was

that the Australian F-111Cs were eventually to be the only service-delivered aircraft not subjected to individual structural testing.

The first F-111C flew in July 1968, with the other 23 closely spaced behind it on the Fort Worth line. What happened next was odd. After being test-flown by GD, USAF and RAAF pilots, it was accepted by the Australian government at a formal ceremony in September 1968. It was then returned to the US Air Force and, followed at intervals by the other 23, taken to pieces and put into storage. The fuselages were parked on their own landing gears at Carswell AFB, the wings were stacked in the GD plant, and the avionics (which were identical with those of the A) were stored in a constant-temperature hangar. While the air and ground crews trained in Australia — over and over again with nothing but engineering manuals, films and systems rigs — the future of the F-111C force depended on the outcome of the USAF structural testing. Every time a wing box cracked it made headlines down under, and in July 1969 Defence Minister Fairhall said that 'until there has been a complete technical evaluation of the results of the tests just concluded it will not be possible to make any detailed comment as to the bearing of these tests on the future of Australia's aircraft'. The usually forthright Aussies seemed to be suffering from a severe case of indecision. Meanwhile, the price kept rising. In 1966 the Australian Auditor-General gave his estimate of the total price as not US$90.75 million but US$216m. In 1968 Fairhall said 'It could cost

Above: Two shots of the first vic of three C-models to nose across the Australian coast at Brisbane, on 1 June 1973, after an 18-hour flight from Nellis. / *RAAF*

Right: Aussie champers for Air Commodore David Evans, CO of the 3,000-strong Amberley base (largest in the RAAF), after his first F-111C solo. Like many base commanders he considers it desirable to keep in touch by doing — who wouldn't, given the chance? / *RAAF*

US$300m', and in 1970 the price had reached about US$320m.

In July 1969 the Aussies, after months of harassment by their national media, decided to send a powerful delegation for what were announced as 'Top Level Talks' (with initial capitals). The team included the Chief of Air Staff, Air Marshal Sir Alister Murdoch; the Secretaries of the Departments of Defence and Air, Sir Henry Bland and F. J. Green; Air Vice-Marshal Hey (whom we met earlier); and the Chief Defence Scientist, Arthur Wills. For many weeks previously a technical team from the RAAF and Department of Supply had been studying the fatigue testing in the United States and wondering what kind of message to send to their government. Now the Top Level Talks were coming to take a decision, and the decision was that the Australians would accept their 24 aircraft and pay an extra $A1.365m each. The Prime Minister, John Gorton, said that, even though the fatigue problem had not been solved, the US government had agreed to 'replace

at no cost for materials or labour' any box that could not be guaranteed for a 15-year life 'at the appropriate time and at no increase in the ceiling price formula'. Just what the last bit meant was unclear, because the original $90,749,040 had been a 'ceiling price' back in 1963. At the same time, the bill for the conversion of six aircraft to the reconnaissance configuration was given as $A30.6m which seemed a bit of a jump from $A7.98m. Labor opposition deputy leader, Lance Barnard, said that if his party was returned at the October 1969 General Election it would seek to renegotiate the deal and buy a 'cheaper, more suitable' alternative.

While the politicians shouted in Canberra, in November 1969 the aircrew at Amberley were delighted to receive their $A4.5m F-111C simulator, made by General Precision/Link at Binghamton, New York, where some RAAF personnel had spent two years. A heavy responsibility lay upon Sqn Ldr B. A. Johnson, OC Simulator, and Flt Lt Tom Carlyon, OC Simulator Maintenance, in keeping the great installation fully and correctly employed so far from the manufacturer and in such a completely adverse political environment. In November 1969 no F-111C had arrived at Amberley and none seemed imminent, whereas the training programme for the base had been planned on flight operations from July 1968. The only Aussie with plenty of F-111 flying in his log book was Flt Lt Ian Westmore, who in 1968-69 had logged over 400 hours with the USAF on exchange posting. In fact, the first team of 50 pilots and navigators had

Above: Then CO of the F-111C force, Gp Capt (now Air Commodore) 'Jake' Newham is seen leaving his aircraft in October 1973 after saluting HM The Queen and the Sydney Opera House opening ceremony. / *RAAF*

Left: Wg Cdr Geoff Talbot, then CO of 1 Sqn, taxies on to the flight-line at RAAF Butterworth, Malaysia, leading a six-plane contingent to the Five-Power Integrated Air-Defence System. This 1974 trip was the first outside Australia by the F-111C. / *RAAF*

Right: Every RAAF crew flies occasionally to Mach 2.2 at high altitude in the 72.5° configuration and with clean wings. / *RAAF*

begun training on the F-111A at Nellis in March 1969, but had had to break off and go home when the first wing box cracked in fatigue testing.

On 23 September 1969 Prime Minister Gorton said that the RAAF would wait for completion of the testing of the revised wing box, which was scheduled to start on 9 October and be completed by 16 November. But an extraordinary series of problems and delays held back the completion of USAF testing until late May 1970. Further long talks ensued, and after conferring with Dr John L. McLucas, Deputy Secretary of the USAF, Defence Minister Malcolm Fraser said he had requested the USAF to 'reactivate the F-111C force as soon as possible so that the aircraft can be delivered'. Fraser said that to delay until mid-November 1970 — the earliest date for delivery following analysis of the testing of the new box — would 'involve a major disruption of the RAAF's programme for flying training'. In any case, he added, the USAF had no plan to rebuild the wing boxes of its F-111A fleet, but would incorporate modifications only on an IRAN (Inspect and Repair As Necessary) basis from 1972 onwards. So at last it was 'all systems go', and the chaps at Amberley eagerly looked forward to getting into the air. The ferry crews left Australia in December 1969, some for the third time.

It was just before Christmas 1969 that a Nellis F-111A had a wing come off, resulting in grounding of the whole F-111 inventory. Fraser issued a press release that was hardly coherent — 'All aspects of the matter will be canvassed . . . The purpose of the visit is to canvass the possibilities concerning the F-111C' — and for the first time mentioned 'alternatives to the present F-111C'. (Indeed at this time the US Secretary of Defense, Melvin Laird, himself said 'If we are going to be plagued with a continuation of these problems we must explore other alternatives'.) So in April 1970 Fraser was talking to the Americans (again), accompanied by the new CAS Air Marshal Colin Hannah; the new Defence Department Secretary, Sir Arthur Tangey; and (for about the 19th time) Air Vice-Marshal Hey. The upshot was that Fraser reported 'The F-111C is at the moment quite unacceptable to Australia and to the RAAF'. The hastily reassembled aircraft were equally hastily re-dismantled and put back into long-term storage. For the first time Australia secured a right to cancel without losing the whole of its F-111 investment, which at that time (without one aircraft delivered) was $A227m. Provided the F-111 was still in the US inventory and the deal was agreed by the US Congress, Australia could cancel and sell the F-111Cs back to the US for a price between $A120m and $A140m. Thus, immediate cancellation would result in a reduced loss of about $A107m (US$86.7m), this sum rising week by week from May 1970 onwards.

Meanwhile, to give the RAAF something to fly, 24 F-4E Phantoms were leased from the US Government at a price of US$35m for two years and US$12m a year thereafter. All 24 were flown to Amberley in 1970, and if the F-111C had been finally cancelled a further 16 would have been bought or leased plus eight RF-4E reconnaissance aircraft and flight-refuelling Buddy packs. The Phantoms gave the Aussies experience of a fairly modern aircraft with all-weather

capability, and took the pressure off the F-111 troubles. In fact the versatile fighter-bomber proved so popular in the RAAF that a strong cell grew up that resisted sending the 23 survivors (one was lost off the Queensland coast) back when the time came.

In December 1970 Opposition leader Gough Whitlam stung Fraser into saying 'Mr Whitlam's call for a decision on the F-111 ignores the facts . . . In April this year I negotiated with the US Secretary of Defense safeguards that were not present in the original agreement: the F-111C must meet RAAF operational and technical requirements; the US will buy them back if they do not; and we need not make a decision until static and fatigue testing is completed late next year . . .' But Lance Barnard attacked the government for the open-ended and 'willy nilly' nature of the original contract, which gave Australia no safeguards of any kind. The theme was even taken up by Fraser himself when he blamed his own Liberal-Country predecessors, Menzies and Townley, for the terms of the agreement.

While all this heat was being generated, the F-111C programme quietly continued, with no announcement that the aircraft would, after all, be delivered. Each aircraft was virtually rebuilt, with 240 chief modifications including completely new wing boxes. The USAF had required one of its own new F-111A

Left: The arrival at Amberley of two USAF KC-135 Stratotankers in October 1976 enabled crews to practise in-flight refuelling during Exercise Kangaroo II. This picture was taken by the right-seater of the next F-111C in the queue. / *RAAF*

Bottom left: This is how an F-111C looks to the USAF boom operator. The receptacle is immediately behind the crew escape module. Pilot of the F-111 on this occasion was Grp Capt (now Air Commodore) Bill Collings, Air Staff Officer at Amberley. / *RAAF*

Right: Armourers of Amberley's No 482 Maintenance Sqn, Leading Aircraftmen Jim MacAnally (driving) and Ken Lawrence (steadying the 500lb retarded bomb) sweat to load 24 bombs aboard each of four aircraft for a firepower demo at the Joint Services Staff College held at Puckapunyal bombing range in Victoria. Mixed in with the load were lo-drag bombs without the tail airbrakes. / *RAAF*

Below: A normal load for a C is four of these pylons each carrying six bombs (in this case retarded 500-pounders). The bomb bay usually carries extra fuel. / *RAAF*

wings, complete from tip to tip, to be fatigue tested to an equivalent of 40,000 hours. The RAAF had asked for only 24,000, but in early 1973 an F-111C wing, after rebuild, was tested to 32,000 hours. In January 1973 a party of 12 aircrew flew to the United States to resume training after a gap of almost four years, followed by further teams in March and May. On 14 March 1973 Barnard, by now Defence Minister in Whitlam's new Labor government, announced how appalled he had been by the 'ineptitude and indecision' of his predecessors, but concluded he had 'no legal or economic alternative other than to see out the contract'. (This was ill-advised; there was no justification for destroying public confidence in the nation's main military aircraft merely to discredit political rivals.) Total cost was then put at US$324m, of which $236m had been spent, and negotiations were started to try to lop $30m of even that total (but in fact it continued to rise, to cover further modifications and provide continuing support). And on 15 March the Qantas shops at Sydney at last began to work on Litton inertial hardware five years after the airline constructed special facilities to handle the F-111 navigation system.

Finally 'F-111 Day' arrived on 1 June 1973. Almost everyone who was anybody in Australia seemed to be in front of Hangar 364 at Amberley to welcome in the first six aircraft, all fully crewed and led by Grp Capt 'Jake' Newham, CO of 82 Bomber Wing. They had been working up at Nellis for five months, and finally took four days (18 flight hours) to ferry from McClellan AFB via Hawaii (Hickam AFB) and Pago Pago AFB in American Samoa. Further sextets flew in, in July, September and December 1973 while the 23 Phantoms were reluctantly flown back to the USA in December 1972 and June 1973. The final six F-111Cs were delayed by shortage of fuel at Pago Pago, had to reposition at Fiji and finally arrived at

Top left: Heat haze from the afterburners of the 18,500lb P-3 engines blots out the trees as a C lifts off. / *RAAF*

Centre left: Like Aardvarks everywhere, 1 and 6 Sqns fly at night. Here the afterburners glow as a training mission departs Amberley into the dusk. / *RAAF*

Bottom left: In this high-lift configuration, with little fuel and no external load, this C-model could fly comfortably at 100 knots. An aircraft with fixed highly-swept or delta wing would have its nose high in the air. / *RAAF*

Above: Afterburner torching of dumped fuel has been used on spectacular fly-pasts by the RAAF to mark the opening of Brisbane's Warana Festival and Melbourne's Moomba Festival. / *RAAF*

Amberley in early December. The RAAF said they were a week late, but in fact they were more than five years late!

On 20 October 1973 nine aircraft from 1 and 6 Sqns, of 82 Wing, flew in formation over Sydney Opera House as the latter was opened by HM The Queen. At last the F-111C seemed to be in business, and this was just the start of a long, tough and thoroughly successful service career. The first major exercises were held in the Darwin area in February and March 1974, in the course of which 1 Sqn flew what I believe were the first sorties ever made by F-111s in the air intercept role; I should think they were also the last. Another notable mission was the 12½ hour flight round Australia via Darwin, Pearce, Edinburgh Field and Pt Cook on 8-9 April 1974, the 50th anniversary of the first such flight (by a Fairey IIID seaplane that took 44 days).

In December 1974 the RAAF awarded GD a contract worth about $280,000 for 'a feasibility study to determine the work required to modify four F-111s for the reconnaissance role.' Just what had been left out of the earlier USAF-funded $118m programme is hard to discover, but Mr Barnard appeared to start from Square 1 as though a reconnaissance F-111 was a new idea! In 1975 the CAS, Air Marshal Rowland, said that four RF aircraft was inadequate in time of serious conflict, but that having four conversions would at least provide a nucleus on which to build. In late 1977, three years after the GD contract, the reconnaissance pod was said to be 'still under development in the US'. Presumably it is not the same as the pod that was fully developed for the RF-111A ten years earlier at a cost of $118m. As the USAF never used these they ought to be going cheap.

In 1975 three C-models completed a three-week exercise off Hawaii with the US Navy. In 1976 two set speed records in high-altitude missions between major Australian cities, flying much further than the direct distances to keep beyond sonic-boom range over the sea; the times were: Adelaide/Melbourne, 27min 8sec, and Sydney/Melbourne 35min exactly, for a distance of almost 1,000km (621 miles). In 1977 intensive exercises were held in the Philippines. Back near Amberley on 28 April 1977, after nearly four years and 21,000 hours, the RAAF lost an aircraft when Capt W. H. 'Bill' Baker and Flt Lt David Clarkson rode back to earth in their capsule after their 6 Sqn machine suffered 'a massive explosion' in the engine compartment. Baker, a Vietnam F-111 veteran on exchange posting from the USAF, evened an old score because an Aussie pilot had written off one of the Nellis aircraft five years previously. The media reported the loss soberly, and noted that the RAAF had expected at 21,000 hours to have lost two aircraft and be half-way to No 3. This can fairly be taken to mean that the F-111C has arrived, won its spurs and been finally accepted as a good piece of kit — which, of course, it had been all along. To clinch this opinion, the Australians are getting an ex-USAF F-111 to fill the vacant slot on the 6 Sqn flight-line.

E for England

Whereas the original TFX intention had been to make as many aircraft as possible with the greatest commonality, the reverse became true. While the numbers were sharply reduced, changes were introduced partly because the basic aircraft was developed for different purposes and partly to rectify what had to be recognised as deficiencies. One of the petulant cries of Senator McClellan's probe was that there should be 'a constructive decision to stop production' until the supposed faults were put right. In the author's experience there has never been such a decision in the history of military aviation except in the case of aircraft dangerous to fly (which, despite the early engine/inlet problems, the F-111 most certainly was not). Thus, it is hard to calculate meaningful costs and consequences of so drastic a decision. Senator McClellan does not represent Texas and so did not appear concerned at the effect such a decision would have had on the 26,500 people at GD Fort Worth, but it should have been self-evident that, in the long term, it does not save money to stop production of an established product either for six days or six months.

One of the new missions foreseen for the F-111 was strategic bombing, and the FB-111 described in the next chapter was an important project from an early point in the programme. Another was the adoption of completely different avionics of the later-technology digital microelectronic type, and this eventually resulted in the F-111D described on pages 92-98. The designation of the D shows that it was meant to follow the B and C, but the development problems with its 'black boxes' took so long to solve that in 1968 the Department of Defense decided to do at least something to improve the aircraft actually being built, and quickly schemed an interim machine called F-111E. In most essentials it was identical to the A,

Right: Aircraft 67-0122, one of the first E-models, pictured while briefly assigned to Cannon AFB before flying to the 20th TFW in England. / *USAF*

but it did incorporate the Triple Plow II inlet. Critics of the programme laid great store by this inlet system, of 14sq in greater area and separated from the fuselage wall by a wider boundary-layer gap discharging through a large grille flush with the dorsal surface, but in fact in nearly all flight regimes it makes no significant difference to performance or to engine behaviour. All it does for the E is improve the performance envelope very slightly at extreme altitude. In theory it allows Mach number to reach 2.5 instead of 2.2, but in practice this is negated by the total-temperature limits on the airframe. At Mach numbers above 2.2 a temperature warning is illuminated on the main caution panel, and a clock counts down from 300 at a speed varying with excess Mach over 2.2, and the pilot has to slow down long before zero is reached or he may hazard the aircraft skin, canopy and other areas. The big inlet also allows more powerful engines to be installed, but as these could not be afforded it is not a very effective debating point.

Apart from the new inlets, the main changes in the E are minor avionic alterations including improvement in the ECM fit (see Appendix). One of the few poor features of the original aircraft, the manual radar-scope camera, was replaced by an automatic strike camera. Easy to introduce, the E followed the A and C on the Fort Worth line in early 1969, and 94 were delivered. By this time the probable future pattern of Air Force deployment of the One-Eleven had become clearer. Instead of having 1,473 identical aircraft

Above left: Take-off from the maker's plant at Fort Worth by a new F-111E on acceptance test. / *GD*

Left: An F-111E has its HF shunt aerial standing out clearly against the ochre paint of the fin. / *GD*

Above: This F-111E has its HF aerial painted over in the normal way. It has yet to receive unit markings. / *GD*

equipping about 20 combat wings, the inescapable answer seemed to be to have a very small number of wings each equipped with a different type of One-Eleven. The 141 production A-models (discounting 18 prototypes) equipped the 474th TFW at Nellis and have been intensively used for training and development, with operational capability in emergency but never in fact assigned an operational role except in SE Asia. The E equips the 20th TFW. As related later the D and F each equip one wing, and the FB strategic bomber is deployed in even lesser numerical strength but equips two wings by making each wing smaller than normal. At least this scheme means that each wing and its base has a common type of aircraft, though it is very far removed from what the Air Force originally intended. The limited number built of each sub-type not only increased costs but also progressively threatens to cause problems in replacing attrition losses. During the useful life of the One-Eleven it is probable that wings will have to accept 'odd man out' aircraft of different sub-types as replacements, and this has hit the Australians already.

This problem has not affected the 20th TFW, because the 20th's operating units, the 55th, 77th and 79th TFS, each have a nominal strength of 24 aircraft, and 72 as a wing total leaves a healthy margin under the total E production of 94. Unlike all previous F-111 units the 20th TFW did not convert in the United States. For 25 years it has been based in England, and it was to RAF Station Upper Heyford, north of Oxford, that the first two E-models to join the Air Force flew on 12 September 1970. The author knew this base during World War 2, flying in Wellingtons as a young cadet. By 1970 it had changed vastly and reverberated to the afterburners of the 20th's F-100s. To look after the much larger and more complex F-111E a building programme costing upwards of $20 million was in hand. Major changes and extensions were being made to the maintenance, overhaul and repair/modification facilities of the 20th Combat Support Group. Complete buildings were being filled with F-111E systems rigs, one of the largest being occupied by the simulator. Each of the three squadrons was receiving new crew facilities, parking pads, quick-reaction alert areas, underground munitions facilities and some degree of hardening against nuclear or conventional attack. Modern aircraft of the calibre of the One-Eleven demand ready access to a major base of this kind, even though in time of crisis they might be dispersed to many airstrips to lessen their vulnerability to a pre-emptive strike.

Not the least significant fact about the 20th TFW is that it is not based in Germany. Details of what its proposed targets might be are, of course, classified information, but the fact it can reach all of them from central England underscores the exceptional range of its big swing-wing delivery vehicle. There is no doubt that the attack radius, with any chosen profile other than Lo throughout, is far greater than that of any heavy bomber operating with full bomb load from Britain in World War 2. Again, however, details of actual radii for given bomb loads are classified information.

Re-equipment of the 20th was completed by mid-1971. By this time the wide spaces of the Oxfordshire base were littered with distinctive Aardvark shapes, uniformly camouflaged apart from the colour of the rear-warning detector fairing atop the fin, which is blue for the 55th TFS, red for the 77th and yellow for the 79th. At first the tail letters were JR for the 79th, JS for the 55th and JT for the 77th, but this soon changed to UH denoting the base. The 20th TFW is, of course, part of US Air Forces Europe (USAFE), headquartered at Ramstein AB in Germany but more locally controlled by HQ 3rd Air Force at RAF Mildenhall, Suffolk. It was at once recognised as by a wide margin the most powerful attack force anywhere in Western Europe, with blind first-pass capability matched only by the RAF Jaguar squadrons which cannot equal the F-111E at night or in bad weather because they lack TFR and attack radar. Until the European-developed Tornado enters service in 1979 the 20th TFW and its partners of the 48th (Chapter 10) bear a giant responsibility in providing the only all-weather 24-hours-a-day attack capability throughout the NATO European nations.

It has been the author's privilege to know many of the 20th's personnel over the years. They are, despite their peaceful environment, as much in the front line as it is possible to be. There are about 3,600 Air Force men and women assigned to Upper Heyford, and there can be few military forces of this size that exert a greater influence on world affairs. In its early Aardvark years the emphasis tended to be on maintaining a low profile, and avoiding publicity — not because the 20th was not proud of its achievements but because the Pentagon was incredibly sensitive about the aircraft and tended to regard any mention of it in the media as bad. Though

Left: Aircrew of the 77th TFS collect flight schedules from the squadron operations desk at RAF Upper Heyford. / *USAF*

Below left: Preflight check of his F-111E by Capt Jim Steiber of the 77th. He is inspecting the dispenser which at the practice range drops up to six small bomblets that hit with a shotgun-shell marking charge so that accuracy can be assessed. / *USAF*

Right: Putting a spanner in the works, an airman from the 20th Field Maintenance Sqn works on a P-3 from one of the wing's F-111Es. When installed, the engine nozzle is completely uncowled. / *USAF*

Below: RAF Greenham Common, where 'Ike' Eisenhower watched paratroopers leave on D-day and which houses Britain's biggest annual airshow, was home to the 20th in the summer of '76 while Upper Heyford's main runway was being resurfaced. / *USAF*

understandable, this sad state of affairs had to die a natural death, and today a generation is growing up that knows little of the Washington-based controversy and rightly regards the F-111E as not only a great technical achievement but also a cornerstone of the defences of Western Europe. Certainly the record of the 20th with this aircraft is impressive. In mid-1971 it was an aircraft of the 20th that chalked up the 100,000th flight hour for all One-Elevens. Today the wing has itself exceeded this total, yet it has lost only three out of its original 72 aircraft — at least two to birdstrikes — and has had no fatalities. In the last calendar year before these words were written (1976) there was not a single major accident. Further emphasising a good record, in 1976 the three Aircraft Maintenance Squadrons comprising the 20th Combat Support Group were selected as 'the best within USAFE'.

The role of the 20th TFW is, of course, long-range 'deterrence' within the NATO Alliance. Air combat is conspicuously absent from the unit's planning, and the general idea is to avoid interception. Modern interceptors armed with snap-down missiles are increasingly acquiring the ability to destroy ground-hugging attack aircraft, but the sophisticated ECM of the F-111E still give it a considerable degree of immunity. At full throttle a clean Aardvark can also show a clean pair of heels to almost any modern fighter. The brochure maximum Mach number at sea level is 1.2, and on occasion aircraft of the 20th demonstrate that this can actually be accomplished. It carries no structural temperature limit and with tremendous internal fuel capacity can be sustained for a long period. Most modern fighters are limited to Mach numbers of 0.9 or slightly higher, and would have no hope of catching a clean One-Eleven skiing at full throttle. The last thing a One-Eleven on a combat mission wants is to engage in a dogfight, and it is strange that most still carry the M61 gun inside their weaponbays instead of removing it and its drum of ammunition to make room for two 750lb bombs. Hardly any crews in operational squadrons have ever fired the gun, and it plays no part in normal mission training.

For the 20th the carefully planned training missions may take crews to almost any part of Europe. Each flight is designed to extend and demonstrate the skills of the crew in simulating — as far as possible — difficult missions in a sophisticated electronic environment. Among the practice bombing ranges are those in Luce Bay (SW Scotland), Jurby (Isle of Man), The Wash, Sardinia and Aviano (not far from Venice). Invariably a practice bomb pod is carried on one of the inboard wing stations, but many missions are scored by radar and camera evidence. Clearly, some of the most important features of a training mission are the electronic performance of the environment and the aircraft. Some of the defensive avionic systems are listed in the Appendix. In Europe it is possible to test these, and the TFR, under almost all conceivable weather conditions.

One of the photographs in this book shows Australian C-models 'torching' dumped fuel. It is possible to pump fuel overboard through a dump nozzle between the jetpipes. If this is done when full afterburner is in use, from either engine, the dumped

Above left: On 3 July 1976 rain came to Upper Heyford as a refreshing change after weeks of scorching drought. Nobody minded, even though it was the 20th's Bicentennial NATO Open Day. / *USAF*

Above: Spirit of '76, Upper Heyford F-111E No 028 in Bicentennial livery to celebrate 200 years of independent friendship. / *USAF*

fuel ignites, burning with a flame which fills the space from 10 to about 60 feet behind the tail. In the picture this is being done for effect, but in fact it is an obvious potential IR defence system, since it provides a source of heat on to which an IR-homing missile might lock in preference to the aircraft jetpipes. With luck a warhead aimed at the burning fuel plume would detonate beyond lethal range, unless it belonged to a large missile of exceptional capability. Of course, the defensive kit also includes IR flares intended to have the same effect, but it is becoming increasingly difficult to fool modern defences which today can filter out flares and even torched fuel and home on the IR emissions from cooler sources such as the hot metal of an engine nozzle.

Nowhere is the situation tougher than in Europe, where — as in North Vietnam — potential enemies have astronomic quantities of radars, communications, computers, Triple-A and SAMs. In 1974 the CO of the 20th TFW was Col (now Brig-Gen) James H. Ahmann. He then told the author:

'War games tend to throw up appalling losses to the attacking air force. However, I do not believe the real world, at this moment, is as bad as this. The defences have distractions, logistics problems, unserviceability — and not everyone wants to win the VC or the Medal of Honor . . . There's a lot of R&D been done that cuts the odds against the aircraft; and the cost/effectiveness analyst is the first guy to get fired when the real shooting starts. In my view the Lo penetration is at present the best, and with the capabilities we have in the F-111 I have no doubt whatever that we could penetrate anywhere we might have to go and hit our targets accurately'.

In December 1977, after this chapter was written, the publisher kindly invited me to a weekend seminar in the course of which participants revisited the 20th TFW. Visible progress was being made with the semi-hardened shelter programme, and with many other items of permanent infrastructure including a large new base hospital. The F-111E was seen with the Westinghouse ALQ-119 tactical jammer pod on the belly rack (aft of the landing gear, not used for weapons), and these pods were painted not the usual white but black. Red stencilled outlines around the canopy of most aircraft denote replacement of the original glass by more flexible multi-layer material better able to withstand severe birdstrikes. Many aircraft formerly without guns have had the M61 put back, though not for air-to-air use and not as a routine mission item. Sadly, the wing had lately lost an aircraft in a headlong plunge into a Welsh mountain, and beyond near-certainty that it was not flying in the TFR mode the loss is likely to remain a mystery. A day or two later, following further losses in other wings, the F-111 force throughout the USAF was grounded, though the restriction was expected to be fairly brief.

The B-52 Replacement

In the late 1940s the long-range missions of the USAF Strategic Air Command could not be flown by any kind of jet aircraft. SAC did buy a large number of the six-jet B-47, but this relied heavily on flight-refuelling and still could not fly the most important long-haul missions. Not until Pratt & Whitney developed a high-compression turbojet of reduced specific fuel consumption, the J57, did it become just possible for a jet bomber to meet SAC's needs. Even then the B-52 had to be an aircraft of tremendous size, powered by eight of the powerful J57s, to carry a nuclear weapon over the required distance. Gradually the technology advanced, and the range of the B-52 was more than doubled, from 5,000 miles to more than 12,000. But these ranges were for missions flown in the thin air at high altitude. By 1960 improvements in defending radars and SAMs had made the high-altitude bomber — no matter how good its ECM — extremely vulnerable, especially one as large as a B-52. Its radar cross-section, a measure of the ease with which it can be detected by the defences, exceeds that for all other combat aircraft in history with the exception of one Soviet type. The B-52 was thus forced to operate at low level, but it was not designed for this arduous duty and the vast airframes suffered terribly from cracks, fatigue failures and fatal crashes.

Another SAC bomber was the totally different B-58 Hustler. The world's first supersonic bomber, this was an even greater technical achievement, and demonstrated such incredible performances as a sustained speed of Mach 2 and range on internal fuel of 5,125 miles as early as 1957. It was a compact, sharp-edged and heavy machine, with extremely advanced flight characteristics bearing close kinship with a One-Eleven with its wings locked at the maximum sweep of 72.5°. Such a situation would be a dangerous emergency for the swing-winger, but to B-58 crews it was routine — and not even difficult, provided that everthing worked. But the B-58 was a demanding aircraft, that still could not fly the long SAC missions, and its basic philosophy and systems had by the late 1960s become somewhat dated. Allied with extremely high costs, this resulted in the decision to withdraw the two Bomb Wings of SAC equipped with the B-58 in January 1970. Today more than half the total B-52 force has also been withdrawn. Out of 744 aircraft built in eight main sub-types only the final two models, the very long-range G and H, remain, a total of 238 aircraft in 17 SAC wings. In each case the aircraft withdrawn have been replaced by the FB-111A, another ill-judged designation which incorrectly suggests close kinship with the F-111A.

The possibility of using a version of the F-111 as a strategic bomber and missile platform for SAC had been discussed since the early years of the TFX studies but these had centred around significantly 'stretched' aircraft with weight increased from the 55,000lb level up to about 100,000lb. The fact that the basic tactical aircraft escalated in weight to the 100,000lb level, with maximum conventional ordnance, tended to kill any prospects for major stretching except by turning the F-111A into a completely different and larger aircraft with more powerful engines. There was even a study for a three-engined SAC bomber with the centre engine fed by a dorsal S-duct. At the same time, part of the escalation in F-111A weight had been due to increased internal fuel capacity to recover some of the range lost through excess drag. This in turn kept the idea of a SAC bomber version alive, and with McNamara's intense personal identification with the programme, and wish to see more versions with maximum commonality, it was always on the cards that it would come to pass.

McNamara announced the FB in December 1965, but before continuing with this important version it is necessary to take a rather large detour through another variant, the F-111K (for the very good reason that most of today's FBs started life as Ks). The F-111K was born to meet a British order, and it was a time of unique ineptitude, emotion and short-sightedness in British aviation which resulted in decisions and non-decisions that did only harm to the F-111 programme and a great deal more harm to Britain. The trouble stemmed from the belief — or, at least, feigned belief — of the British Labour Party that the all-British TSR.2 aircraft was a thoroughly undesirable and costly mistake that should be done away with at the earliest possible date. Had the

Right: US Air Force No 63-9783 was originally the 16th F-111A development aircraft. Rebuilt as the first FB-111, it carried out most of the basic flight programmes associated with this version apart from SRAM compatibility. / *GD*

USAF
U.S. AIR FORCE
39783

Labour government, which took office in October 1964, stated that Britain had no need of such an aircraft it might have had a case. Instead, the view was expressed that TSR.2 should be cancelled and replaced by the American F-111. As the two aircraft were strictly comparable in missions, capability and timing, the arguments had to centre on the belief that, while the American aircraft was a scintillating marvel, the British one was a 'can of worms' that had no chance of succeeding.

The root cause of the vendetta against TSR.2 was really the vague belief that, in building the aircraft a succession of inept governments had asked for, the British planemakers had taken the public for a ride and wasted millions on unwanted prototypes. In fact, the industry had been almost driven round the bend by foolish defence policies and continually changed decisions, and a Conservative Defence Minister had announced in 1957 that there would be no more manned combat aircraft at all! This incredible policy made TSR.2 almost impossible to start, and in 1965 it stood out like a sore thumb as the nation's only major new warplane. Two smaller and less-advanced programmes were cancelled almost as soon as the Labour government took office, the Hawker P.1154 being replaced by the American Phantom and the HS.681 V/STOL transport being replaced by the American C-130. The TSR.2 was allowed to survive until April 1965, but by this time — contrary to all the government's hopes — flight development was going so well that it could have made cancellation more difficult. A technical paper before the Royal Aeronautical Society on the flight-test programme was therefore suppressed, the F-111 extolled to the public as a far better and much cheaper alternative, and the British aircraft cancelled on 6 April 1965. Numerous impressive figures were then published to show how much would be saved by buying the American aircraft.

Above: Another view of the FB aerodynamic prototype showing a dry hook-up with a tanker whilst carrying three dummy SRAMs. This aircraft was not delivered to SAC. / *GD*

Top right: A January 1971 picture taken during the acceptance flight of one of the first FB-111As for SAC. Clean wings at about 45° sweep look graceful. / *GD*

Centre right: Hardly visible, a SAC FB streaks at full throttle with wings at full sweep. This is the only USAF version with long-span wings. / *USAF*

Bottom right: Nose of an FB-111A bomber opened up to show the main elements of the APQ-114 multi-mode radar, with the two small dish aerials of the APQ-110 terrain-following radar visible under the main dish. All the GE radars used in F-111 versions other than the F-111D look similar. / *General Electric*

True to British style there was no searching public examination to discover the truth, but merely wild statements by politicians of all parties who knew little but expressed their views with vehemence. Thus, while Prime Minister Wilson said each F-111 would cost 'less than half the estimated total' for TSR.2, his Aviation Minister (Jenkins) later said the price of 'the Mk 1 F-111 is 20 per cent less than the price at which we did our comparative calculations with the TSR.2'; and while most Ministers in April 1965 gave £300m as the 'saving' from buying the F-111, Defence Minister Healey later put the total saving of all three axed

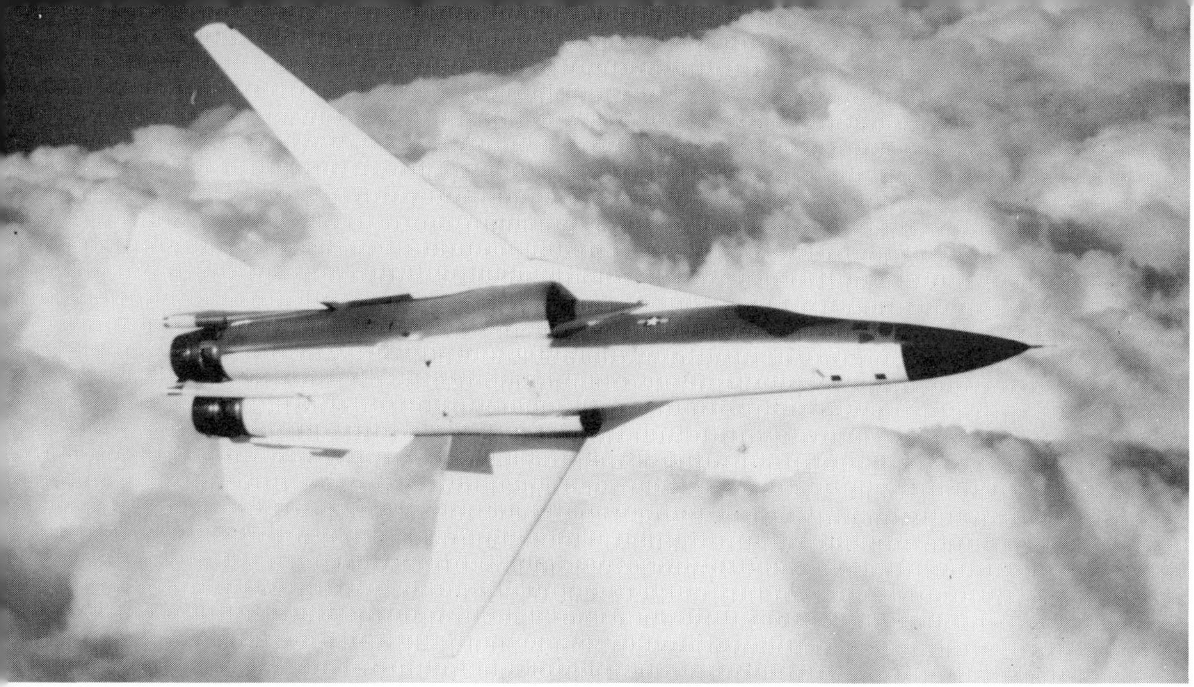

programmes as only £110m. Actual prices were hard to discover, but in December 1965 the Plowden *Report on the Aircraft Industry* commissioned by the government gave the unit cost of TSR.2 as £5.5m and that of the F-111 as 'hardly one half of this'.

Such naive imprecision helped, because, had the British F-111 buy gone through, the 'ceiling price' of $5.95m would certainly have been multiplied by more than two, making nonsense of the much-vaunted cost savings. It would probably have quadrupled, because not only would Britain have had to pay a share of a much larger total R&D bill but the aircraft selected for the RAF would have had the 'Mk II' avionics which later proved to cost more than four times the original estimate. On 1 February 1967 Defence Minister Healey announced that the RAF would receive 50 F-111K aircraft, costing in all £135m through the first five years of operation. (Three months later the cost for 15 years, which one might have expected to be £155m on Healey's basis of £2½m per aircraft plus £2m a year operating cost, was announced as £336m, without a word of explanation.) Though based on the F-111A, the K was to have the long-span wing of the B, most of the advanced Mk II avionics (apparently all supplied from the USA), the strengthened main gears and increased-capacity brakes and tyres. John Stonehouse, then Minister of State for Technology, visited Fort Worth and said the greatest thing about the F-111 was its ferry range 'which means it can be deployed throughout the United Kingdom'. Nobody ever did discover what he meant to say, and in any case the only UK base was to be Cottesmore, which had previously been equipping for TSR.2. There were countless further nonsenses and contradictions,

FB-111A 53 & ON

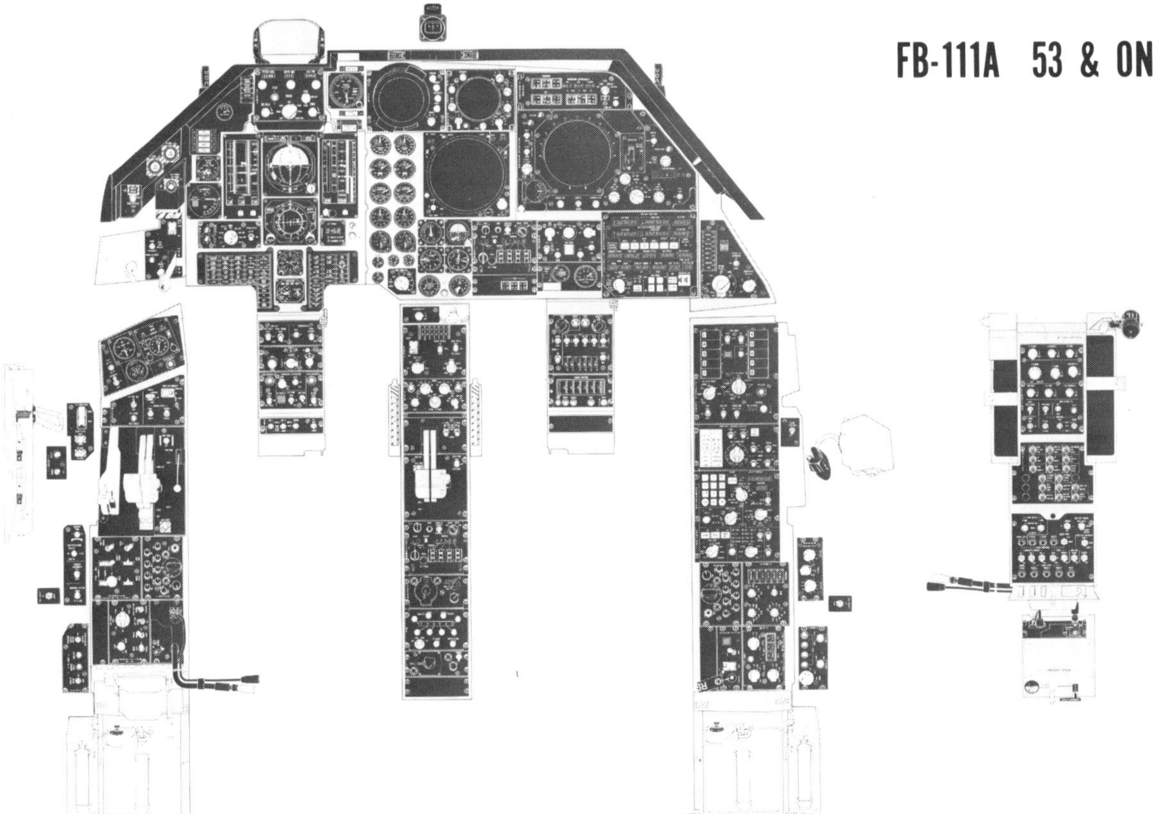

Top left: FB-111A serial 67-0161, pictured after delivery. Soon it wore the star-spangled blue band of SAC. / *GD*

Bottom left: Basic cockpit panels in the FB-111A from No 53 on. These later aircraft have an additional electronic display for the navigator and rearranged consoles. / *USAF*

Above: One of the most costly 'Link trainers', the FB-111A flight simulator is a product of the same company that made the first 'blue box' more than 40 years ago. The instructor consoles are off to the left. / *GD*

including 21 weeks of pointless discussions on the single topic of the UK wish to be quoted 'a fixed price' and McNamara's explanation that he did not have legal authority to do such a thing. By December 1966 this point was reinforced when Healey gave the cost for a mere nine years as £425m. A month later Prime Minister Wilson announced cancellation, claiming it would 'save £400m'. Pity Britain could not go on forever ordering aircraft and cancelling them, because by saving £300m by ordering the F-111 and saving another £400m by cancelling it the nation seemed to be on to a good thing.

To return to reality, GD had to try to build the 50 K models, even though its customer (which was the US government) had no idea of the price and did not even have a final build-standard from the British Ministry of Defence. By January 1968 all 50 had been through detail manufacture, 19 were taking shape on the assembly line and the two YF-111K flight-test aircraft were virtually complete. These were oddballs, too far gone to be turned into a USAF service type, and they were redesignated YF-111A and used for various USAF research programmes (one of them was passed to the National Aeronautics and Space Administration and modified at NASA's Flight Research Center at Edwards with supercritical swing-wings, with which valuable aerodynamic research at various sweep angles was performed in 1972-75 in a programme called TACT, Transonic AirCraft Technology).

The other 48 F-111K airframes were used in the FB-111A programme, several being already assembled and merely completed to the FB standard. Structurally the K and FB would have been identical, but the SAC bomber has different systems and equipment. When he announced the FB programme at a press conference on 10 December 1965 McNamara said the new aircraft would have 'twice the speed [of the early B-52 models], with approximately the same range'. In fact, the speed comparison depends greatly on whether one is considering low altitude or high; at low level the clean FB-111 is less than twice as fast as a good B-52 but at high altitude it is about three times as fast. On the other hand, when carrying maximum weapon load it is markedly slowed down, though still faster than the old monster. McNamara said the cost of the 263 aircraft needed to maintain the authorised SAC force of 210 aircraft, including spares, would be 'on the order of a billion and three-quarters' ($1,750m).

Though powered by a more advanced TF30 engine, rated at over 20,000lb thrust, the FB has to carry more fuel and sometimes a heavier weapon load than the A-model and usually has a thrust/weight ratio even lower than the original tactical machine. This is no great handicap, and unlike the TAC aircraft the SAC bomber invariably can count on being able to operate from a 10,000-foot concrete runway. The long-span swing-wings are restressed, and though they carry only six pylons the permitted external loads can substantially exceed those of the TAC versions, at the penalty of remaining subsonic, not doing TFR skiing and keeping within restrictive *g* limits. With the wings not swept beyond 26° and (exceptionally) fitted with eight pylons it is possible to carry 50 Mk 117 bombs of nominal 750lb (two internally and six on each pylon), and as the actual weight of these is about 825lb the maximum bomb load is a remarkable 41,250lb. With this load on board the gross weight exceeds 122,000lb, and at Military (maximum non-afterburning) power the ceiling is described as 'lower than Pike's Peak' (a mountain of 14,110ft). This is of little significance. What is much more important is that the FB-111A was about 30 per cent down on range, though actual figures have not been published.

While each FB crew, comprising pilot and navigator (he is styled thus, though he is at least as much a bombardier and defensive-systems operator), do on occasion fly a maximum-weight mission, most training sorties are flown at lower weights. With internal fuel only, the FB weighs about 81,537lb, and with six SRAM missiles about 95,100lb. With four 500 Imp gal external tanks the weight is 100,458lb, and with six rather over 110,000lb. A typical load in a nuclear conflict might be six Mk 43, Mk 57 or TX-61 free-fall bombs, two of them in the internal bay. Against strong defences SRAM would be by far the most important weapon. Weighing about 2,230lb, it is an almost perfectly streamlined missile with three tail fins set at 120° but without wings. A pulsed rocket motor propels it at highly supersonic speed on any desired trajectory, at high or low altitudes, with options of skiing over the ground at tree-top height, taking evasive action or hitting its target from behind. Though its initials signify Short-Range Attack Missile its maximum range at high altitudes is over 100 miles. Despite its small diameter, and tiny radar cross-section, it carries a formidable thermonuclear warhead. The FB-111A carries a computer and other special avionics to make it compatible with the slim white missiles, with which it could destroy six major targets — cities even — in a single sortie.

The first FB, a converted F-111A, flew on 30 July 1967, followed on 13 July 1968 by the first production aircraft (fitted temporarily with P-3 engines). The first delivery was made to SAC's 340th Bomb Group at Carswell AFB on 25 September 1969, where the Commanding General, Bruce K. Holloway (who as USAF Director of Operational Requirements had played a central role in the early years of the TFX programme) ceremonially accepted it on 8 October. He said the FB would 'play a substantial role in contributing to the SAC mission of deterrence through the 1975 time period'. At that time there seemed every chance the B-1 would soon become the prime manned delivery system of SAC, but this was not to be. The FB, and the creaking old BUFFs (Big Ugly Fat Fellas, as the B-52 is affectionately known), will have to soldier on not just to 1975 but possibly to 1995, or whenever the wings finally part company with the fuselage.

What makes SAC's task harder is that range is, as noted earlier, 30 per cent short. When he became SecDef Melvin Laird cast a critical eye over the FB-111A programme, and on 19 March 1969 he said it did not meet the requirements for an intercontinental bomber. Laird was SecDef for a new administration, and might have been expected to knock the F-111. He cut procurement from 263 to a mere 76, 'to salvage work in progress'. McClellan gleefully reported that these 76 would cost $1.2 billion, or 70 per cent of the planned cost of the original 263.

Below: The first true FB-111A pictured during SRAM trials at Eglin AFB. White dots on the aircraft and pylons aid in tracking with precision cameras and theodolites. / *GD*

Below right: A production FB-111A with wings at 45° seen immediately after releasing a SRAM, before missile motor ignition. In this case the missiles are also painted with camera target dots. / *USAF*

In service this force supports 60 aircraft in the combat inventory, as listed in the Appendix. For eight years they have operated on average more intensively than other USAF combat aircraft, in arduous circumstances, with alert forces at ready status around the clock not only at the two home bases but also dispersed to an average of six airfields at any one time. In May 1970 SAC participated in the RAF bombing competition, and two of the brand-new FBs were allowed to go along not to compete but just as a navigational exercise. One finished just pipped into second place by a B-52 after having led the field for most of the contest. In November 1970 the swing-wing bomber officially participated in SAC's own competition. Two crews entered, and one came first in bombing and the other second in bombing and first in combined bombing and navigation.

In most respects the FB is a valuable delivery system. When the author met the SRAM Program Director, Col Lawrence A. Skantze, at an early stage in the missile's introduction to the Air Force, he was left in no doubt that it is regarded highly. According to Skantze, 'Compatibility between the airplane and missile so far has been excellent . . . In fact, it is better than in the B-52, and the superior inbuilt navigation system of the FB-111 indicates that we will consistently achieve better CEPs with that aircraft'. CEP, rendered as Circular Error Probable or Circle of Equal Probabilities, is the circle within which half the hits may be expected to fall when aiming at a point target, or the circle within which the probability of any one missile hitting is 50 per cent. With SRAM the CEP is always within the effective radius of the warhead, but when launched from an FB-111A the area of CEP is roughly halved.

To close this chapter we have to take a brief look at how SAC shapes up in the wake of the cancellation of the B-1 in July 1977. Though the Soviet Union was insistent that the FB-111A should be discussed in all SALT (Strategic Arms Limitation Treaty) negotiations it has never been regarded as a true strategic bomber; it is not in the same class as the much larger and heavier Tupolev *Backfire*. Yet opponents of the costly B-1 programme, recognising the vital need to maintain a force of manned delivery systems (without which the United States could be faced with the terrifying decision either to start a nuclear war on the mere radar detection of an enemy attack or else wait until it was destroyed), called for a development of the FB-111 to be bought instead of the completely new B-1. Unfortunately the two delivery systems are as alike as chalk and cheese, and the probability that the FB will eventually carry the ALCM (Air-Launched Cruise Missile), either AGM-86A or Tomahawk, does not alter the fact.

The proposal was summed up in a letter to Senator Barry Goldwater, dated 23 February 1976, by the Commander of SAC, Gen Russell E. Dougherty 'This alternative has the initial appeal of offering a more modern and higher-performance penetrator, since the

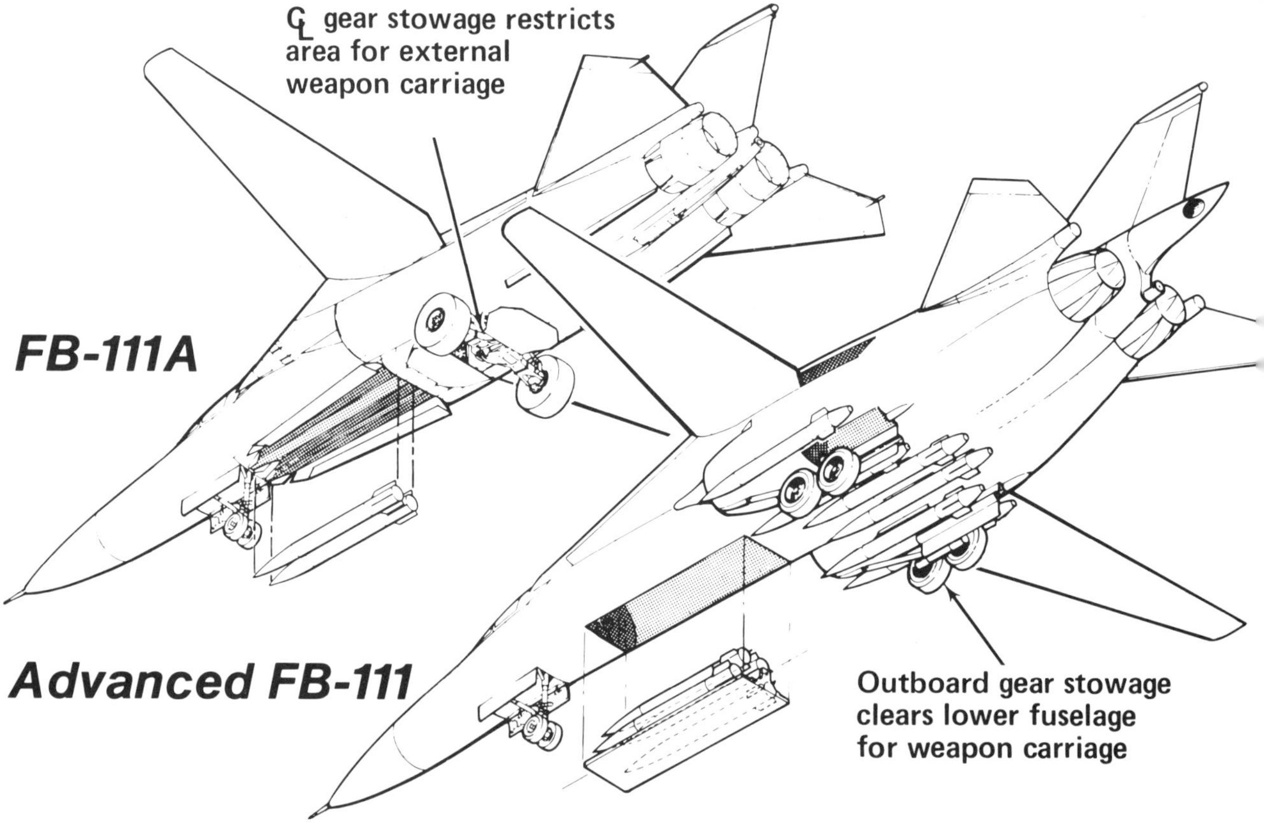

FB-111 is basically a hard and fast aircraft with low radar reflectivity. However, our continuing analysis of the various proposals for FB-111 upgrade has led to the conclusion that the extensive modification required to make the FB-111 comparable to the B-1 would be, in effect, an entirely new aircraft. The basic design . . . does not have the growth potential to compete efficiently . . . It does not have the desired low-level range and payload characteristics, and, in order to do the job required, would have to be procured, manned and supported in such large numbers that it is neither an economical nor efficient alternative to the proposed B-1 force. As respected as the FB-111 is within SAC's manned bomber force, we have a pragmatic recognition of its limitations in size and range, neither of which can adequately be overcome by modification.'

This was a fair assessment, but after President Carter cancelled the B-1 the US Air Force was placed in a critical position. How could it maintain a credible manned delivery system throughout the 1980s? Certainly not with the B-52 nor with the FB-111A, but there was a possibility that Congress would agree to fund a stretched FB-111. Systems Command dusted off the earlier studies and by August 1977 had

Above: Crews like the existing main landing gear, which extends as one unit (so it is impossible to get one wheel down and one up). But it has the grave fault of eliminating the possibility of a heavy load of fuselage-hung weapons. In the proposed FB-111H there would be a new weapons-bay pallet carrying four SRAMs on a rotary door, and new tandem-wheel gears allowing six more SRAMs or other weapons to be hung on new fuselage pylons. / *GD*

Right: This most informative series of drawings shows the main features of the proposed FB-111H. It would have a new, longer fuselage with much more fuel, new weapons bay and different landing gear retracting outboard of the changed inlet ducts to the more powerful but shorter engines. ECS is environmental control system and APU auxiliary power unit. /*GD*

crystallised its ideas which became an increasingly open secret (though as this book is written, a month after that time, still nothing has been said officially in public). There are several options, but the most important and likely action is procurement of about 80 completely new aircraft to a substantially different design and probably to be designated B-2A.

The chief modifications would be to the fuselage, which would be considerably enlarged and at least 10ft longer. It would have internal tankage for a total of approximately 9,000gal of fuel, weighing nearly 72,000lb and increasing maximum takeoff weight to about 160,000lb. Engines would be two General Electric F101 advanced augmented turbofans almost identical to the engine cleared for production of the B-1, each rated at 17,000lb dry and over 30,000 with full afterburner. The weapon bay would be enlarged to accommodate a rack launcher for four SRAMs, while each of the six wing pylons would carry a side-by-side pair of cruise missiles. Unofficial reports claim the new aircraft would have 'almost the same range as the B-1', but this is not possible with maximum weapon load when a practical radius of action of 2,500 miles would appear the limit.

Prior to this new-build programme there there would certainly be at least one rebuild of an FB-111A, which might fly in 1980. R&D cost has been estimated at $380m, plus $195 million for further development. There is also a chance about 65 of the existing FB-111A force might be similarly rebuilt in 1980-83 at a cost of about 2.3 billion 1977 dollars. The new programme of 80 additional aircraft would probably cost $4.2 billion. How far the new or rebuilt aircraft could equal the instant-reaction, nuclear-resistant capability of the B-1, or the latter's immense defensive-electronics systems, remains a matter for conjecture.

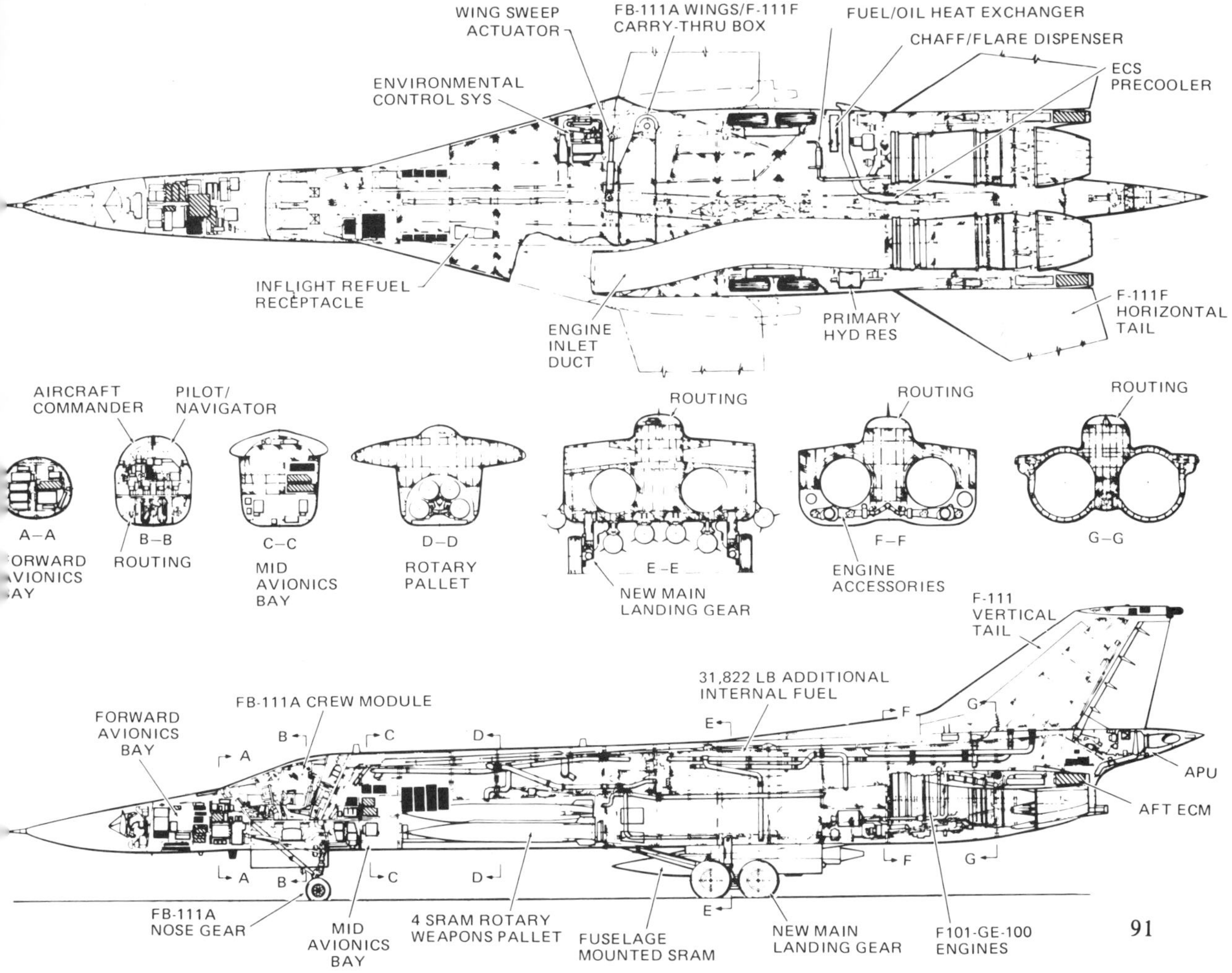

Mainly about Technology

Every aircraft in history has been a compromise. Increasingly, the design teams are faced with the need to choose the right level of technology, and these decisions are not only superimposed on the basic decisions of aerodynamics, structures and propulsion but interact with them to make the process fraught with difficulties. To some degree the whole point of the TFX was that it was new technology, with swing-wings, afterburning turbofan engines, rough-field landing gear, TFR and much else. Yet many of the reasons for selecting the GD aircraft were because it used what was thought to be older technology than its rival. It had no thrust reversers, dorsal inlets or titanium primary structure. Its avionics were 'state of the art' black boxes from which the new technology of microelectronics was generally absent. True to their style, Pratt & Whitney designed the TF30 to be reliable rather than small, light and clever.

It was sheer bad luck that many of the parts of the original F-111A that were specially planned to avoid trouble — such as the propulsion system and the wing box — presented some of the biggest problems. But, even if they had worked perfectly, their deliberately conservative design made them candidates for eventual modification or replacement, either in new sub-types of One-Eleven or by retrofitting to the original model. This was true in many of the functioning systems, but by far the most important area for eventual updating was the avionics. In the F-111A these systems had been far more comprehensive than in any previous tactical aircraft, but they were heavy, filled the whole of the long Aardvark nose, and turned a lot of electrical power into unwanted heat. (For list, see Appendix.)

Microelectronics promised several advantages. This radical new technology had been visible ever since the

Right: An F-model in TFR flight in the area of the Rocky Mountains.

MO
366

first operating transistor in 1948. By the late 1960s transistors and related 'solid-state' devices were almost universal in avionics, finally allowing the elimination of almost all the unreliable and bulky thermionic vacuum tubes; but they were still supplied as 'discrete devices' which were picked up by the fingers or tweezers and crimped or soldered on to a printed-circuit board. The resulting board might be about 3 x 2 inches, and hundreds would be attached by multi-pin edge connectors to a hinged 'mother board' at least a square foot in area, of which there would be several in the final 'black box'. With microelectronics all the devices, with few exceptions, are created inside microscopic layers of single-crystal silicon, or similar material with a precise repeating crystal lattice structure, by knocking out atoms of one element and replacing them with atoms of another. The technique is obviously on a microscopic scale, and the only way to make what are called 'microcircuits' is by reducing everything photographically until the circuit and its devices can just be seen under a microscope. The final silicon wafer is scribed into chips about as large as the head of a pin, and these are then mounted on something big enough for tweezers to grip and coupled up electrically. The result is black boxes which are not only almost embarrassingly small — it would be the simplest thing in the world to fit a complete radio inside its own on/off button or volume control knob — but which consume hardly any power and almost never go wrong. Reducing power consumption in turn means that less heat has to be dissipated, and the overall result is a dramatically better electronic system. The only items where high power is still needed are radars, jammers and similar units where sheer power is essential to do a large job over ranges of many miles.

Several US aerospace companies were making microelectronic units by 1960, one of the first being

Above: The same aircraft carrying 24 bombs in TFR flight over the Rocky Mountains.

Above right: Though outwardly difficult to distinguish from other tactical One-Elevens, the F-111F has almost 30 per cent more engine thrust — and outstanding avionics. / *GD*

the guidance package for the Minuteman I missile. The prime contractor for this was North American (now Rockwell International) Autonetics, and it was this company which, with a growing team of suppliers, discussed the proposed TFX 'Mk II' avionics in 1961-63 (the description 'Mk II' was a usage even the British had abandoned). By 1963 a complete Mk II specification had been agreed, but it presented a few challenges and the intention to incorporate the new avionics at the 100th aircraft in 1968 was regarded as tentative. Autonetics received a systems-integration contract valued at an estimated $60m, on top of the contracts for hardware.

In a very few cases the Air Force decided to let well alone, and this proved a wise decision. The most important was the TFR, a system vital to the F-111 mission and one which had performed outstandingly well. Early 'breadboard' testing was so promising the Air Force decided not to include it in the Mk II package, and this was confirmed by performance in testbeds in the early 1960s and in the first TFR-equipped F-111A in November 1966. Texas Instruments and Sperry merely introduced microelectronic circuits into the same basic system packages, without changing functional parameters. But the attack radar was completely new. Autonetics itself had for years been working on microelectronic-circuit systems for advanced airborne radars, and it secured the contract for the APQ-130 which promised exceptional versatility. In the Lo TFR mode it can map the ground ahead with photographic sharpness,

despite the low grazing angles, providing MTI (Moving-Target Identification), better identification of all targets and better air-to-ground weapon delivery. In the air-to-air mode it can detect and track targets even in the presence of heavy clutter, heavy rain or other adverse conditions, and simultaneously emit a narrow CW (Continuous-Wave) beam for guiding air-to-air missiles such as Sparrow.

Allied with the APQ-130, and integrated with the TFR, radar altimeters, flight-control system and navigation systems, are a completely new computer and cockpit display. The new computer is one of the System/4 Pi family by IBM, one of a very advanced all-digital, all-solid-state series which are tailored for each application by assembling the selected sub-packages, each hardened to withstand adverse military environments. The Mk II avionics includes two of these computers, one for navigation and the other for weapons delivery, installed in a complex of much greater versatility, memory and processing speed than the analog complex in the F-111A. In the cockpit the pilot and WSO can both see an IDS/HUD (Integrated Display Set and Head-Up Display) giving two impressive optical displays combining radar and other information. An HSD (Horizontal Situation Display) gives aircraft position, heading, track and destination (waypoint or target) superimposed on navigation maps, reconnaissance photographs or pictures obtained by radar or IR linescan.

Clearly, the Mk II integrated avionics added up to the greatest package that could be devised for tactical aircraft in the early 1960s, and the chief thing wrong with it is that it took too long to develop and (partly for this reason) cost too much. Reluctantly, the Air Force decided not to retrofit it to F-111A aircraft already built, but it did expect the F-111D, the new model designed to carry it, to become the standard type as soon as the system was fully developed. But the first F-111D, a rebuilt A, did not fly until 2 December 1968 (long after the new system was meant to be on the production line), and the production D did not emerge until the autumn of 1970 at No 236 on the production line. By this time the R&D bill, estimated at $60m, had considerably exceeded $280m (it reached that level in early 1969), and to avoid making too much of a loss the price of each Mk II installation had reached $2.2m. Procurement of the F-111D was therefore cut to 96, equipping only a single wing, the 27th TFW which it reached in October 1971.

Long before this time the Air Force had gone through an agonising rethink and decided to abandon the costly Mk II system and instead buy an assemblage of systems that offered most of the capability for much less money. This was a major win for GE, who had lost the main radar to Autonetics and now got the business back again. The company's Aircraft Equipment Division in Utica first produced an improved radar for the FB-111A and then, as the main element in the Mk IIB avionics system, developed a more significantly altered radar designated APQ-144 for the final model of F-111, the F. Combining the best features of the earlier -113 and -114, the -144 has several 'performance enhancement' features, of which the most important are a brief 0.2 microsecond pulsewidth and an extra-short display range of down to 2.5 miles. The Mk IIB avionics makes no pretence at air-to-air modes, which in the F-111D were merely wasted cost and effort. GE have flown DMTI (Digital Moving-Target Indication) in a -144 and several other advanced features, but so far as the author is aware these have not been retrofitted in the F-111F. The F also carries many other advanced equipments, notably in the field of defensive avionics, as listed in the Appendix. The computer complex remains the System/4 Pi.

Both these improved versions of the F-111 have later engines than the somewhat primitive and troublesome TF30 in the F-111A. It will be recalled that the interim F-111E introduced the larger Triple Plow II inlet system, but retained the original P-3 engine. In the F-111D the engine was changed to the P-9, which owed something to the earlier P-12 which had been developed under intense pressure to provide increased thrust for the F-111B. But by the late 1960s Pratt & Whitney, in various programmes, had learned how to make significantly better engines and with the JTF10A-32C it added more than 25 per cent to the thrust, with greater reliability, better handling and reduced fuel consumption. This vastly improved engine was chosen for the F-111F and went into production as the TF30-P-100. Among its many new features (for the pre-1970 era) are: new fan and LP compressor with bulged inner profile giving canted blades handling greater airflow with no flow separation; completely new Finwall combustion-chamber cans, which need half as much cooling air, cut weight and give smokeless combustion combined with higher gas temperature; completely new HP turbine with directionally solidified cast blades and film-cooled stator vanes operating at 1,260°C (compared with 1,120° for the P-3 and 1,140° for the P-9); and a dramatically improved afterburner and blow-in ejector nozzle.

This new engine transformed the F-111F into a wholly outstanding aircraft which could, if it wished, behave as quite a respectable fighter as well as a bomber. It is still not in the same class as an F-15, because it is older in concept and much heavier, but its engines actually put out more thrust. For the record, the ratio of thrust to weight (T/W) in the F-111A is 0.39 while that of the F-111F is almost 0.53, a very great improvement. Yet one should not read too much into these figures. Though the F is much more sprightly than the A or any other One-Eleven, the older versions can fly their missions perfectly well, and that is really all that matters.

Another advance in the F is the incorporation of boron-epoxy composite material. These new materials are formed by embedding hair-like filaments of boron (formed by depositing boron vapour on a white-hot tungsten filament) in an epoxy-resin matrix. Like the British-pioneered graphite composities, they are extremely light, strong and stiff, and they were first tested in F-111 components, such as wing panels and landing-gear doors, in 1967. In December 1970 the decision was taken to incorporate boron-epoxy doubler plates into the production line, and the obvious place to do it was on the vexed wing-pivot fitting which had earlier caused the loss of an aircraft. The first swing-wings actually to receive this reinforcement, which reduces stress concentrations and thus greatly extends fatigue life, were those of the Australian C-models, which were largely rebuilt before delivery in 1973. Every F-111F had the doubler from the start, and it is the intention of the Air Force to add it to earlier aircraft as they come in for IRAN (Inspect and Repair As Necessary).

Predictably, the only thing wrong with the F-111F is its cost. Despite the fact that the Mk IIB avionics cost little more than half as much as the Mk II system, general price escalation and reduced production rates

Above left: Most F-111 models are powered by the TF30-P-3 afterburning turbofan, rated with maximum augmentation at 18,500lb. Behind the fixed inlet vanes can be seen the first-stage fan rotor blades which in 1965 caused such stall problems. These engines were not designed to contain fan blades that broke off, but in 1978 an improved version (for the F-14A Tomcat) can do this.
/ *Pratt & Whitney*

Above: Outwardly almost identical to the P-3 engine, the TF30-P-100 is far more powerful, with a maximum rating of 25,100lb. / *Pratt & Whitney*

forced up the price with each successive model of the Aardvark, instead of the hoped-for gradual reduction resulting from the favourable effects of the 'learning curve' (a mathematical plot based on the fact that, in normal production, each item can be made more easily and quickly than the one before). Companies are sometimes coy about prices, unless they have to argue publicly in Washington, but an idea of what the F-111F cost is provided by the fact that the planned force of 219 had to be cut to only 82 because of escalating costs, and to pay for these the Air Force voted $566m in 1970 and $283m in 1971 — a total of $849m, which according to the estimates at the beginning of the F-111 programme would have bought 350 aircraft. Subsequently a further series of large votes approved by Congress sufficed to keep open the production line at a low rate, and the manufacture of new aircraft finally closed in early 1976 with the 106th F-111F and the 562nd overall. According to the 1974 Secretary of the Air Force, Richard C. Seamans, the overall unit cost of the One-Eleven throughout the programme, including everything, was 'about $15 million'.

Delivery of the F-111F began in February 1972 to the 347th TFW at Mountain Home AFB, Idaho. Until the 41st aircraft the engines were still of the P-9 type, but by 1973 the P-100 was installed and it has since been retrofitted in all the F force. In theory it would be possible to convert earlier engines to P-100 standard, but it would cost more than the new price of the P-3 or P-9 engines and has not been attempted. In 1973 the 347th was assigned to the SEA theatre under PACAF, operating the F-111A from Takhli and other bases in Thailand until 1976. The F-111Fs were transferred to the 366th TFW, still at Mountain Home, from where one squadron was temporarily detached in August 1976 to Taegu AB, Republic of Korea, to reinforce a friendly country that once again appeared to be under threat of invasion (nothing happened, perhaps because of the USAF presence). Aircraft arrived at Taegu 11hr 30min after departure from Mountain Home.

In March 1977 the F force changed hands yet again, flying the Atlantic non-stop on internal fuel to arrive at RAF Lakenheath, England. Here they replaced the F-4D Phantoms of the 48th TFW, the total number of inventory aircraft having risen with the final batches to 84, with approximately 20 in store for attrition. Today the 48th can fairly claim to be the most powerful offensive unit of any kind in the European NATO nations. With their friendly rivals in the 20th at Upper Heyford they are the only outfit that could hit any target in Europe, any time, any weather, even on the 4th of July.

Fighting with Electrons

A recurrent theme throughout this book is that, despite its 'F' designation, the mission of the F-111 is to penetrate hostile airspace and hit surface targets. It can do this at least as well as any other aircraft in the world, and better than most (some would claim, better than all). It has so far needed no help from other aircraft to fly its missions, but, like other aircraft before it, it will soon need a little help if it is to survive. Other tactical aircraft need help already, and many would have little chance of surviving even the outward part of a mission without electronic support. Most tactical aircraft carry ECM pods and other 'penaids', but these are not enough.

Penaids, short for penetration aids, comprise a great wealth of devices that may be carried in or on the attacking aircraft or in special support aircraft that fly the same mission either with the attacker(s) or just beforehand. One obvious penaid is the TJS (Tactical Jamming System) that broadcasts extremely powerful signals on the same wavelengths as enemy radars, communications channels or SAM guidance frequencies. Others are chaff (described previously), flares (likewise) and even decoys in the form of miniature RPVs (Remotely Piloted Vehicles) or other winged or powered devices that try to look like the attackers to the enemy radars. Almost all modern tactical aircraft carry ECM (Electronic CounterMeasures) pods which include noise jammers and often other devices, as well as various warning receivers that light up or give a shrill alarm when triggered by hostile radars, fighters or SAMs. But for really well defended airspace this is not enough. Small external pods lack the potential to jam with enough power; enemy surface radars may have instantaneous signal powers measured in megawatts (millions of watts), and can hop from one frequency to another in a seemingly random way. What chance do puny ECM pods and small add-on systems have against such power and anti-jamming cleverness?

Listen to Lt-Gen Robert C. Mathis, Vice-Commander, Air Force Systems Command, speaking at the unveiling of the first true EF-111A on 20 May 1977:

'We have only to reflect on the Vietnam era to realise the impact of the enemy's capability to detect attacking air forces. With long-range radars of Soviet design and manufacture the North Vietnamese were able to detect our strike aircraft from gear-up to their attack run. The element of surprise was completely denied. The Soviets have taken the impact of electronic warfare and made it the keystone of their defensive forces. The East European environment today reveals the most dense environment of early-warning and ground-control-intercept radar networks known to exist in the world'.

For a tactical attack aircraft to carry enough to help it through such an environment it would have to leave off all its bombs, its weapon-delivery systems and all other mission equipment and turn itself into a purpose-designed EW (Electronic Warfare) aircraft. This is precisely what has had to be done, in the past, and now the past has caught up with the One-Eleven. Virtually all existing tactical EW aircraft are conversions of former attack machines. The first of real importance was the Douglas B-66 Destroyer, which during the 1960s was recognised as being rather large and slow for survival in hostile environments but to have the size, electrical power and space needed for comprehensive EW systems. The EB-66 in several sub-types has for the past 15 years been the most important ECM support aircraft of TAC, and it accompanied most offensive missions by USAF aircraft in the SEA theatre. After 1969 it was supplemented by the EF-105, again in several versions, in completely rebuilt ECM forms carrying passive warning receivers, signal detectors and analysers, jammers, chaff/flare dispensers and recorders. The importance of the EW support mission is shown by the fact that each EF-105 meant the loss to the attacking forces of an F-105, one of the best and most 'penetrable' of all tactical attack aircraft. It also underlines the fact that the EW platform must itself be able to penetrate and survive, and despite its immense built-in protection the EB-66 was finding this increasingly hard to do.

By 1970 the USAF had decided that the F-111A, of which quite large numbers existed in various training

Right: Another salvaged F-111A, in this case a former Nellis inventory machine, mounted in a wood frame at Kirtland AFB for measurement of the nuclear 'hardness' of the EF-111A by subjecting it to giant EMP (electromagnetic pulse) effects. / *USAF*

Above: Take-off at Calverton of the first F-111A modified by Grumman to test the aerodynamic (performance and flight control) effects of the EF-111A canoe radome. / *Grumman*

Left: A salvaged F-111A (the 15th development prototype) mounted on a pedestal at USAF Rome Air Development Center to check the EF-111A aerial radiation patterns. / *Grumman*

Right: In the world's largest anechoic chamber, at Grumman's Calverton establishment, this former Nellis F-111A was used to test the electromagnetic compatibility of the TJS subsystems that now fly in the EF-111A. / *Grumman*

and research roles, was the ideal EW platform for TAC in the future. Unlike all other aircraft it had the speed, range, carrying capacity and internal volume. The Air Force also noted that in the Grumman EA-6B the Navy had an outstanding tactical EW aircraft, far better than anything in TAC's inventory. The EA-6B Prowler is a complete rebuild of the A-6 Intruder, the only attack bomber able to rival the F-111 for all-weather navigation and weapon delivery. Compared with the A-6 the EA-6B has a larger front fuselage housing a crew of four, and an extremely comprehensive TJS, the AIL ALQ-99, plus many other EW items. The chief emitting elements in the ALQ-99 system are carried externally, because there was no room inside the aircraft. Four large underwing pods each house two very powerful TJS transmitters, powered by ram-air turbogenerators on the nose of each pod. Forming the top of the tail is a large fairing covering special receiver aerials (antennas), and there are other jamming, sensing and receiving subsystems throughout the aircraft, as well as a computer control complex and cockpit displays and interfaces with the crew. The whole electronic payload weighs just over three tons.

This battle-proven installation was the obvious base on which the Air Force wished to build its future EW aircraft, and the F-111A was the obvious aircraft in which to put it. What made the marriage especially happy was that Grumman had always been a main associate contractor on the One-Eleven, building aft fuselages even after collapse of the F-111B programme. But there remained to be answered several large questions. Would an EF-111A have to carry a crew of four (which would have cut severely into internal volume and probably made the aircraft subsonic)? Would the ALQ-99 have to be carried externally, cutting performance still further? On the positive side, could the whole TJS be improved and updated, to meet the growing threats posed by potential enemies in the final quarter of the 20th century (because an EF-111A would have an active life of about 25 years)? There were even more basic questions. Would the EF-111A be able to survive in expected 'sophisticated' hostile airspace? Was a large manned aircraft the right vehicle for airborne EW, or should the answer be fleets of small RPVs (either recoverable or expendable)?

Before describing how the EF-111A was developed, it is worth standing a little further back and looking again at the situation in a modern aerial attack. Aircraft can be detected by various means: radars, IR heat sensors, laser beams, sound receivers and the eyes of observers. Similar methods can be used to guide anti-aircraft missiles or Triple-A fire, and the same methods can again be used by the attacking aircraft to hit the target. There is thus no single system that must be used either to ensure a hit on the target or to protect the target against the oncoming aircraft. In the past tactical aircraft have used a diversity of methods, and their protection systems have varied from nothing up to powerful and clever barrage jammers of several kinds, deception jammers that put the defences 'off the scent', passive systems such as chaff and flares (at best confined to one patch of sky for a very limited time-period) and the aid of purpose-built EW support aircraft. It is the vastly increased capability of the support aircraft that make them worthwhile. They alone have the space, electrical power and computer

power to handle the whole of the enemy threat. They have to sense every one of the enemy's emitters (radars, IR sources, lasers and communications transmitters). Within microseconds they must completely analyse the radiation (frequency, pulse length and repetition frequency, and other fundamentals); then they must at once provide the appropriate counter.

The answer may be simply 'barrage jamming', blotting out traffic on the enemy wavelength. In Europe simple one-frequency emissions are likely to be rare in future, and the modern EW aircraft has to cope with 'frequency diversity' in which radars switch from one bandwidth to another, and 'frequency agility' in which one never knows the frequency of the next pulse. Agile radars are clearly difficult to jam, because on a pulse-by-pulse basis it is very hard indeed to stop the required information getting back to the defenders — it calls for a thing called an FTLO (Fast-Tuned Local Oscillator). Thus today's airborne EW installation has to have not only immense electronic power, and high capability in IR and other electromagnetic wavebands, but also enormous EDP (Electronic Data Processing) capability, both to keep on top of the maze of hostile emissions and manage the on-board systems. EW management in the modern attack aircraft is vital if power is not to be wastefully dissipated.

Above: Electrical characteristics of the 16ft 9in canoe radome were measured by mounting it upside down in a special rig at Grumman's original plant at Bethpage (where the Wildcat, Hellcat and Avenger were built). / *Grumman*

Above right: The remarkable EF-111A cockpit mock-up achieved jointly by Grumman and the US Air Force which enables a pilot and EWO (Electronic-Warfare Officer) to manage the whole TJS. / *Grumman*

Whatever may be the role of the RPV in future EW, or the capabilities of individual future attack aircraft, the US Air Force decided there was a need for an EF-111A. Grumman achieved considerable success in one of the biggest-ever system-management jobs to create this new version which, if put into the TAC inventory, could dramatically improve the survivability of other Aardvarks, the F-4, F-15, F-16, A-7 and A-10. It would have three major roles. In barrier/standoff the TJS orbits along the FEBA (Forward Edge of Battle

Area) just out of range of the enemy's fire, creating an electronic wall and thus greatly reducing warning time of friendly attacks. In the penetration/escort role one or more EFs, with or without RPVs, fly ahead of or alongside the attacking force, neutralising the defences on the outward journey, over the target and on the return. In the CAS (Close Air Support) mission the EF loiters around the battle area jamming and confusing the enemy defences to clear a path in which the CAS attackers can operate almost at will. (Of course, all this presupposes that the vital EF is not hit by crude weapons such as infantry fire, manually laid Triple-A or a command-guidance SAM such as Tigercat or the optically tracked version of Rapier.)

In 1974 GD and Grumman were awarded design study contracts for the EF, and in January 1975 Grumman received an $85.9m contract to design, develop, test and evaluate two F-111As rebuilt to the EF-111A configuration. Though seemingly substantial, the amount of the award was remarkably low in view of the immensity of the task. It is almost open-ended, because there can never be a perfect EF, but always the best compromise possible at any given time. By 1975 Grumman had come to the conclusion that the EF-111A could operate without two additional crew members and with the TJS carried internally, both of which were large advances. The chief element in the TJS, the ALQ-99A, occupies racking space filling the fuselage behind the cockpit and in the weapons bay, and the high-power emitters feed aerials disposed along a 16-foot canoe radome on the underside of the bay. In December 1975 Grumman flew an F-111A with a dummy canoe to ensure that neither performance nor handling of the basic aircraft would be harmed.

On 17 May 1977 the second of the two EF-111A conversions flew at Calverton, complete with all its operating subsystems. The most obvious modification is the fin-top fairing housing the receiving equipment and aerials, this being almost identical to that in the EA-6B. Internally, of course, there are many new systems, as listed in the Appendix and including the ALQ-137 SPS (Self-Protection System) and ALR-62 TTWS (Terminal Threat Warning System). The basic ALQ-99A is in important respects improved beyond the ALQ-99 system of the EA-6B, one obvious feature deducible from the lack of the two human operators being greater automaticity. It clearly has greatly augmented and extended computerised management, to respond to more threats faster, pick up and identify enemy emissions faster and make better use of the available power. Again one can deduce that the power of one million watts is generated on board, probably by engine-driven machinery, for there are no ram-air turbogenerators. Grumman can also comment that the -99A provides 'more independent jamming signals in a wider range of frequencies', which possibly could not be deduced. There is a choice between directional or omni-directed (360° coverage) transmissions, certainly using a greater number of transmitters and possibly with several sending out computer controlled emissions of different kinds whilst fed by a common-signal source. According to Grumman v-p Bob Miller 'four or five EFs flying racetrack patterns in Eastern Europe could screen off all friendly airspace from the Baltic to the Adriatic'. Clearly Grumman have done a great job in packing these powerful and complex systems inside the already dense-packed airframe, and in accommodating the displays and control panels in a pretty busy cockpit.

As this book goes to press, flight trials with the first two EF models appear to be going well. It is the intention of the Air Force to fund 40 EF-111A conversions of the original A-models, to equip two TAC EW squadrons 'in the late 1970s'. The Fiscal Year 1978 DoD budget contained $31.6m to convert two additional aircraft, but this is hardly a full go-ahead on the inventory programme. The truly worrying problem is cost escalation, which could become serious enough to kill the programme. Without the EF it is hard to see how tactical aircraft — of any Western nation — can survive in the 1980s.

Flying the F-111

A pilot or weapon system officer (WSO) coming to the F-111 from another fighter would notice several slight variances at first glance. The sheer size of the airplane is impressive; it grosses out at over 80,000lb with internal fuel only, compared to, say, an RF-4C's 52,000lb with three external fuel tanks. The F-111 may also look somewhat cluttered, what with wing high-lift devices (slats, flaps, rotating gloves) and pylons. The wings are high and set well back; only the outer halves of the wings (which seem to be set very far aft) can be seen from the cockpit. When one first sits in the cockpit the limited visibility is immediately noticeable. Because of the side-by-side seating arrangement the pilot must rely on the WSO to clear the sky visually on the right side, and occasionally cross-cockpit formation flying must be done. EB-66 pilots, who sat on the left and had no right-seaters, will appreciate the problem. This matter is really not terribly important, since the F-111 works mainly alone, at night, on the deck, at flight parameters not approachable at present by other fighters.

A glance down the top of the aircraft will reveal a broad, flattish back, which generates a great deal of lift. A position about 100 yards behind an F-111 will give you a good view of the very wide dorsal area as it gently slopes down toward you. In the air, at forward wing sweeps and with gear down, the F-111 looks like a great mottled goose with long beak and high wings. But sweep the wings full aft and an incredible transformation occurs. The ugly duckling becomes a graceful, deadly, dartlike creature as it knifes through the sea of air — no frills and all business.

Since the F-111 has no ejection seat but employs an escape module, the usual detailed seat-preflight is eliminated. Strapping-in is a simple affair, and evidence of foresight and human engineering in the cockpit is readily apparent. It is a comfortable cockpit, and the airplane has often been referred to as the 'Cadillac of fighters.' There are actually places to stow charts, checklists, and flight publications.

Starting the engines is quite simple — and flexible. In addition to the usual pneumatic start, a starter cartridge can be used, and cross-bleed starts can be accomplished. Hot starts are almost unheard of. Unlike the J79 engines in the F-4, which idled within very small permissible deviations from 66% rpm, the TF30 can idle from 57 to 69% rpm.

Taxiing the F-111 is also a little different. The crew sit directly over the nosewheel, and the large main wheels are fairly close together. When making a tight turn one gets an impression of teetering. On takeoff, nosewheel liftoff averages about 142 knots. At Nellis Air Force Base, Nevada, where base altitude is over 1,800 feet and summer runway temperatures rise to 125°F and higher, takeoff distance is about 4,200 feet. Acceleration is fairly good for such a heavy aircraft. In the early A models maximum engine thrust is around 18,000lb per engine; later models have far more powerful engines in the 25,000lb class. The stick is pulled back at rotation speed, but immediately after nose-liftoff a forward stick motion may be required to arrest rotation of the aircraft. The pilot cannot snatch the gear up, as in other aircraft, because of the exceptionally long extension of the main gear struts. If the gear handle were raised while the wheels were still in contact with the runway, a pressure of 750lb/sq in would be routed to the wheels for auto-braking, and two blown tyres would result.

The variable-sweep wings rotate on individual pivots and literally permit the pilot to redesign the aircraft in flight. The pivots are steel pins 8½inches in diameter, set in a solid box which is the structural heart of the airplane. Wings extended to 16° provide a maximum surface area for greatest lift, permitting short takeoffs and landings. As speed increases and lift is converted to drag, wingspan and surface are reduced by sweeping the wings back to a maximum of 72.5°, where they lie tucked along the tail. No retrimming is necessary while sweeping the wings or changing speed, since the aircraft automatically and consistently trims itself to 1*g* flight. As the wings move, the four inboard pylons also automatically readjust their alignment to remain streamlined at all times. The versatility and overlap of the system was dramatically illustrated by the four-ship 'tiger formation' employed for displays by my old squadron: the leader flew with wings out at 26°, No 2 and No 3 set sweep of 45°, and the slot man was swept full aft at 72.5°. It was a pretty impressive sight as we flew over

Right: Left- and right-seaters of the 390th TFS immediately before closing the canopy and heading out for a practice 'show of strength' mission from Taegu. Note thickness of the pilot checklist. / *USAF Official*

TFS

the field. The aircraft will slip through Mach 1 with about 54° sweep; at Mach 1.7 or thereabouts the wings will be positioned to 72.5°, and the F-111 will punch through Mach 2 and higher. Because of its bulk, the F-111 makes a tremendous supersonic shock wave. Standing in a desert valley bombing range, I have seen the entire valley floor instantly erupt in a cloud of boiling dust, as if sharply struck by the flat of a giant hand, following a low-level supersonic bomb run.

The F-111, although a fighter [*sic,* Bill Gunston], can carry an incredible bomb load. It has an internal weapons bay in addition to the wing-mounted pylons. A wide variety of ordnance can be loaded, including various sizes of general-purpose bombs, CBUs (cluster bomb units), rockets, and air-to-air missiles. A six-barrel 20mm cannon is also installed.

After takeoff it seems to take a long time to clean the airplane up; the gear is raised, then the slats and flaps. Normal cruise wing sweep is 26°. The F-111 handles beautifully in the air. Like the gooney birds of Midway — which appear awkward and ungainly on the ground — it assumes a new personality when airborne, exuding grace and confidence as it glides over the Earth. It is a very honest and responsive aircraft, and many pilots comment on the fact that there is no 'dead' spot in the stick. A constant *g* response is commanded for a given amount of stick-travel regardless of airspeed, altitude, or gross weight.

The range of the F-111 is astounding, especially for a flyer coming from another fighter. F-111s have crossed the Atlantic nonstop and without inflight refuelling, on internal fuel only. At the time of this writing a Nellis crew has just returned from Pease Air

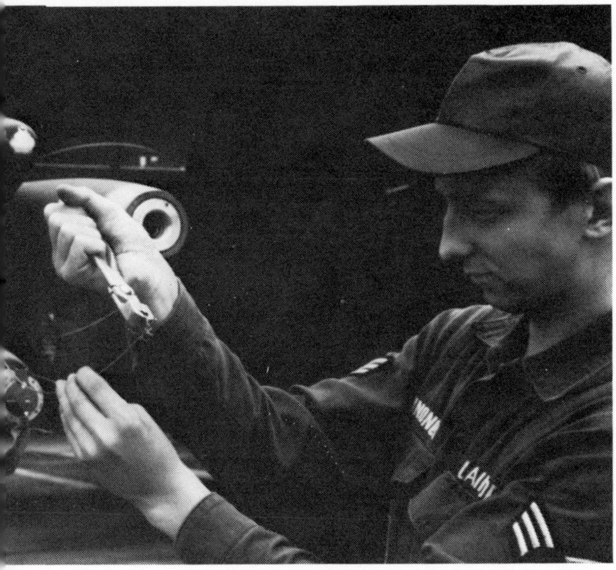

Left: Two pictures showing a preflight on an F-111F shortly after arrival at Taegu AB, South Korea, in August 1976. Carried out by the aircrew, Crew Chief and selected ground tradesmen, the task involves checking numerous external items (illustrated, bomb racks and main landing gear). / *USAF Official*

Bottom left: Final check of the flight-crew information before a talk with the weather staff and a bus ride to the aircraft at Taegu, August 1976. / *USAF Official*

Above: SrA Arthur C. Kinonan wires the fuzes of Mk82 bomb triplets loaded aboard an F-111F at Taegu. / *USAF Official*

Force Base, New Hampshire, a distance of about 2,250 miles. On the trip east they arrived over Pease after five hours and were still able to shoot several approaches and closed patterns before finally landing. This great range is due to another technical innovation in the F-111 — the fan engine with five-stage afterburner.

I'm not sure if anyone really knows what the F-111's top speed really is. On my first flight in the F-111 we reached Mach 2 while climbing out from takeoff, still relatively heavy, and had to reduce speed only to keep from running out of the supersonic corridor. A nice turn was made at 1.8, and the aircraft performed exceedingly well (turns, rolls, etc.) while supersonic. It is designed to provide a stable bombing platform while supersonic on the deck, and it does this very well. The advertised top speed at altitude is Mach 2.5, or about 1,640mph. For sustained high-Mach operation a total-temperature indicator is provided. At very high speeds, when critical skin temperatures are approached, a red warning light comes on and commands the pilot to reduce speed. A digital readout will start to count down from 300 seconds. A 'Total Temperature' lamp on the main caution panel will also light up. The pilot had better pull the power back at this time or structural damage could occur.

Should the crew be forced to abandon the aircraft, the only conscious effort required of either crew member is squeezing the ejection handle. Everything after that is completely automatic. Explosive charges sever the module from the aircraft. The module's rocket motor has lower and upper nozzles, and special sensors continually monitor aircraft speed and direct firing of the proper nozzle. At low speeds the lower, more powerful nozzle is fired, with a slight assist from the upper nozzle to maintain correct trajectory. *G*-sensors and barostatic devices initiate deployment of the recovery parachute. The crew can select automatic chaff initiation at module separation to assist ground radars in detecting the bailout area. Centre of gravity considerations are critical to ensure optimum escape attitude, and the weight differential between the crew members may not exceed 65lb. Even the thermos water bottles stowed above and behind each crew member must be filled and carried aboard, since they have an effect on proper module loading. A Nellis crew bailed out over the rugged terrain near the Grand Canyon not long ago. The module rolled down an incline for several hundred yards. The crew, strapped inside, emerged relatively untouched and stated that they might not have survived had a conventional ejection seat been used. My own experience in bailing out in a conventional rocket ejection seat over similar rugged country, with severe injury to the pilot, confirms this observation. The module will float if it lands in water and extra flotation bags can be inflated; should high swells cause cabin flooding, the aircraft commander's control stick becomes a bilge pump. The crew can also sever the module from the aircraft while under water. Again, flotation bags inflate to help the module rise to the surface.

Have you ever watched a mallard duck come in for a landing? It comes in with neck taut, wings outstretched, and stubby legs close together and braced. That's the landing F-111 in a nutshell. Those who have landed aircraft like the F-101 and F-104 will appreciate the landing speed of the F-111. It will touch down within a few knots of nosewheel liftoff speed — somewhere around 140 knots, depending on landing weight and wing sweep (which will be 16° or 25°). The F-4 with the boundary-layer control system approximates this. Initial approach speeds are standard — 350 knots indicated. Pitchout is made with throttles at idle, and 10-12 alpha units (an angle-of-attack measurement which is simpler to use than airspeed). Rollout on downwind is still at idle power, using what seems like a lengthy procedure to attain landing configuration. The main-gear door is the speed brake. The gear is lowered first, followed by the slats, then flaps. Base turn is made with a bank angle of 30-40°, in contrast to the 50-70° of bank in some other fighters. Halfway through the turn on base, power is usually applied. With the fan engine, which accelerates relatively slowly, the pilot cannot afford to wait until he needs power to apply it in order to arrest sink rate. No matter how tightly he pitches out, once in landing configuration it is always possible to turn inside the pitch.

The final approach is with power on, holding 10 alpha. Ground effect is experienced upon approaching the overrun, which tends to shallow out the angle of attack. This is countered by applying slight back-stick, which is definitely not a flare or roundout. As soon as touchdown is achieved and the engines retarded to idle, wing spoilers rise to kill wing lift and assist in slowing the aircraft. The F-111 has no drag chute, but braking qualities are so excellent that short-field landings are no problem. The tyres are big, and maintenance gets ten times more landings out of these tyres than from those on any other fighters.

To those with experience in high-speed low-level flight, one thing that is instantly noticeable is the steady, granitelike ride. Other fighters flying within similar low-level parameters ride 'hard' — the aircraft is striving hard and the ride is bumpy in any sort of unstable air. The F-111 glides through the Mach at low level so easily that the crew would never know the difference were it not for the Mach indicator. The reason for this is the stability-augmentation system. This flight-control system is unique in that it is self-adaptive. Heretofore, the only other aircraft to possess this feature was the X-15. Triple-redundant pitch, roll, and yaw circuits constantly compensate for deviations in aircraft motion; and computer-commanded corrections are performed before the crew is aware of any possible turbulence or gusting. The result is one of the most solid, stable rides imaginable.

The TFR (terrain-following radar) is another system that makes this airplane unusual. With the deployment of the 474th Tactical Fighter Wing to Southeast Asia in September 1972, the United States finally had a full-time, credible night low-level strike capability. Terrain-following radar systems were not new; *automatic* TFR systems were. The missions flown by the 474th over North Vietnam could not have been accomplished manually. The crew can select a variety of altitudes as well as kinds of ride desired: a three-position switch directs a soft, medium, or hard ride. Soft ride programs a gentle pull-up over high ground; hard ride brings the aircraft closer to terrain before initiating a steeper pull-up; in hard ride the descent commands up to a zero-*g* pushover. The TFR also has a situation mode, which permits the crew to fly around peaks rather than over them, and a ground-mapping mode which does what the name implies. One can also make excellent fixed-angle bomb runs with the ground-mapping mode of the TFR, which provides a backup in case the attack radar system (ARS) is lost. The TFR system is dual-channel; should the operating system fail, another identical system will automatically take over. If a malfunction occurs anywhere in the system, fail-safe circuits will command an immediate and shuddering pull-up.

Both ground and airborne TFR checks are performed to test the radars and the fail-safe circuits. The aircraft will automatically descend on command, hands off, at a rate of about 10,000 feet per minute. At 5,000 feet above ground level the radar altimeter will identify the Earth's surface, and the rate of descent will increase noticeably until just prior to the pre-selected altitude where a smooth level-off is automatically accomplished. The stick will not move as the aircraft climbs and descends over mountain and valley; the climb and dive commands are sent directly from the TFR computer to the pitch channel of the flight-control system. Occasionally it may be necessary to apply power as a particularly steep mountain is approached. In the inky blackness of night it is initially unnerving to look out and see the black hulks of mountains slide by above you as the aircraft settles into a valley.

The WSO backs up the TFR with the big ARS and continuously alerts the pilot to the terrain picture ahead. The ARS has outstanding resolution. Its frequency-agility characteristic does two important things for the operator: by continuously and automatically shifting frequencies it approaches optimum reflectivity of the wide variety of materials swept by the radar and provides an antijamming function as well. After using equipment like this I could never think of returning to other aircraft I've flown — such as the B-52, EB-66, and RF-4C — much as I enjoyed them all.

As the target is approached, the WSO punches a button that recalls target coordinates set into the computer before flight. The radar cross-hairs automatically reach out and settle on the target, and the inertial navigation system is generally so good that the usual problem of last-minute radar scope interpretation is eliminated. If the target is a 'no-show,' offset bombing capability is available.

The TFR system will keep the aircraft at the desired bombing altitude. However, if the BCU (ballistics computer unit) is being used, delivery parameters are not critical; the BCU will constantly and automatically adjust ballistics for changes in heading, airspeed, and altitude until bombs-away. Should the BCU fail, manual ballistics inserted into the system

Above: An F of 366 TFW Detachment 1 takes off from Taegu on its return to Mountain Home at the conclusion of Korean deployment. / *USAF Official*

before flight can be instantly used. F-111 crews still debate whether the auto-bomb mode using the BCU is as accurate as trail bomb using manual ballistics; it's a moot point, since the bombing circular-error averages in both are so good. Visual lay-down nuclear bombing attacks can be made at 200 feet and 520 knots minimum airspeed, using the pilot's lead-computing optical sight.

While aircrews have always been totally mystified at the bad press the F-111 has received, the aircraft finally began to receive some well-earned praise as a result of the Linebacker I and II campaigns — the air offensives over North Vietnam in 1972. Night after night the F-111s took off from Takhli Royal Thai Air Force Base, Thailand, armed with Mk 82 500lb bombs or Mk 84 2,000lb general-purpose bombs, descended over Laos to terrain-following altitudes, flew at low altitudes as far as the northeast railroad, delivered their ordnance, and returned to Laos at low altitudes before climbing out for home. Well over an hour was spent at very low altitudes and high speeds, and the entire mission from start to finish was flown with internal fuel only and no inflight refuelling. No other fighter in the world can even approximate this performance. Medium-altitude missions were flown against enemy forces in Laos and against the supply trails along the panhandle, and the great accuracy of the radar bombing systems repeatedly broke the backs of Communist attacks. The legendary General Vang Pao, the Laotian Lawrence, presented the 474th Tactical Fighter Wing with a trophy AK-47 rifle as an expression of thanks. Captured North Vietnamese troops referred to F-111s as 'whispering death'.

On days when solid undercast would normally have prevented fighter strikes on enemy forces, we rendezvoused with flights of F-4 or A-7 aircraft that released their ordnance on command of the F-111, which was bombing by radar. This technique of pathfinder bombing proved to be very valuable and was employed night and day. While the B-52s rightly earned great praise for their part in the final December blitz over North Vietnam, it was the F-111 aircraft which — singly and largely unheralded — struck the enemy MiG airfields, SAM (surface-to-air missile) and gun sites, and other threats minutes before the big bombers appeared on the scene and directly contributed to minimising B-52 losses.

It is ironic indeed that it was the North Vietnamese themselves who did a great deal to build the legend that is growing around this airplane. Communist air-defence forces never succeeded in solving the problem of detecting intruding F-111s. POWs returning from Hanoi told of Communist target areas sounding the all-clear after strikes by other US aircraft, only to be completely surprised minutes later by lone F-111s roaring over and attacking their targets. One of the Nellis crews, who was hit by a 'golden BB' in December, told of the Communist prison guard who came up to him and said, 'You F-111.' Then making a flat horizontal sweeping motion with the palm of his hand, he shook his head in admiration and said, 'Whoosh.' An amazing compliment by the enemy. I don't think I can improve on that.

Appendix

MODELS AND USERS

YF-111A Two aircraft originally ordered by USAF as the first two F-111K for Britain but completed as RDT&E aircraft for USAF following cancellation of F-111K programme. Included in F-111A total.

F-111A Original tactical model for USAF, with first 18 RDT&E prototypes given same designation as subsequent examples, despite significant differences throughout these and the following 23 production aircraft. Original RDT&E batch had USAF serial 63-9766 to -9783, total 18. Production blocks were numbered 65-5701 to -5710, 66-011 to -058 and 67-032 to -114, total 141. Most of the RDT&E aircraft were assigned to General Dynamics, AF Flight Test Center and other establishments, and were later passed on to systems-manufacturers, NASA and the EF-111A programme. Production aircraft were assigned to 4520th Combat Crew Training Wing, Nellis AFB; 4480th TFW, Nellis AFB; 474th TFW, Nellis AFB; 347th TFW, detached to Takhli RTAFB; today serving only with 366th TFW, Mountain Home AFB, and 57th FWW, McClellan AFB.

F-111B Fleet air defence fighter for US Navy, with P-12 engine, short nose, extended wing-tips and many internal changes, configured primarily for air-to-air missions (unlike all other models); used in RDT&E programmes at Grumman, Hughes, NAS Patuxent River, NAS Miramar, NAS Point Mugu and NAS Lakehurst. Total five prototypes plus two production.

F-111C Tactical aircraft for RAAF based on F-111A but with long-span wing-tips and strengthened landing gear; assigned to 1 and 6 Sqns, RAAF Amberley, Queensland. USAF serial numbers 67-125 to -148, total 24.

F-111D Improved tactical aircraft for USAF with 'Mk II' avionics and P-9 engine; assigned to 27th TFW, Cannon AFB. USAF serials 68-085 to -180, total 96.

F-111E Tactical aircraft for USAF with improved inlet system (to match later TF30 engine, never fitted) and advanced penaids and weapon-management subsystems; assigned to 20th TFW, RAF Upper Heyford, and 57th FWW, McClellan AFB. USAF serial numbers 67-115 to -124 and 68-001 to -084, total 94.

F-111F Tactical aircraft for USAF with 'Mk IIB' avionics and P-100 engine; assigned first to 347th TFW, then transferred to 366th TFW at Mountain Home AFB and now serving with 48th TFW, RAF Lakenheath, and 57th FWW, McClellan AFB. USAF serials 70-2362 to -2419, 71-883 to -894, 72-1441 to -1452, 73-0707 to -0718 and 74-0177 to -0188, total 106.

EF-111A Rebuilds of F-111A aircraft as tactical EW (electronic warfare) platforms carrying jamming and other electronic support systems; intended programme involves conversion of 40 aircraft. Grumman is prime contractor.

FB-111A Bomber and missile platform for USAF Strategic Air Command, with long-span wing-tips, strengthened landing gear, special navigation and SRAM equipment and P-7 engine; assigned to 340th Bomb Group, Carswell AFB; 4007th Combat Crew Training Squadron, Carswell AFB; 4201st Test Sqn, Eglin AFB (SRAM mission); today serving with 380th BW(M) at Plattsburgh AFB (30 aircraft) and 509th BW(M) at Pease AFB (30 aircraft). USAF serials 67-159 to -163, 67-7192 to -7196, 68-239 to -292 and 69-6503 to -6514, total 76.

FB-111H Proposed bomber and missile platform for USAF Strategic Air Command, produced as FB-111A conversion and/or new-build programme but with no decision at end-1977; fuselage stretched 104in and enlarged to accommodate 31,822lb additional fuel (total 64,574lb), new weapon bay with rotary pallet carrying four SRAM or five B61 nuclear bombs (or three B43 or four B77), tandem twin-wheel main gears leaving fuselage underside clear for six SRAM or nuclear bombs, simple fixed-geometry inlets feeding two F101-GE-100 engines (each in 30,000lb thrust class with maximum augmentation) and APU/starter in new tailcone.

SPECIFICATIONS

Dimensions

Span (16° sweep, F-111A, D,E,F,EF) 63ft 0in, (16° sweep, B,C,FB) 70ft 0in, (72.5°, A,D,E,F,EF) 31ft

Above: A contrasting view of an FB at full sweep. This is the heaviest of all F-111 versions. / *USAF*

11.4in, (72.5°, B,C,FB-111A) 33ft 11in, (60.0°, FB-111H) 44ft 10.2in.
Length (without probe, except B and H) 73ft 6in, (with probe, except B and H) 75ft 6¼in, (B) 66ft 9in, (H) 88ft 2.5in.
Height (except B,EF and H) 17ft 1.4in, (B) 16ft 8in, (EF) 21ft 9in, (H) 22ft 0in.

Weights

Empty, equipped (A, typical) 45,200lb, (B) 46,112lb, (C) 47,300lb, (D) 46,900lb, (E) 45,700lb, (F) 47,200lb, (FB-111A) 47,500lb, (FB-111H) 51,832lb, (EF) 53,600lb.
Max take-off (A) 92,500lb, (B) 79,000lb, (C) 114,300lb, (D) 92,500lb, (E) 92,500lb, (F) 100,000lb, (FB-111A) 114,300lb, (FB-111H) 140,000lb, (EF) 87,800lb.
Max after refuelling in flight (A) 100,000lb, (B) as MTO, (C) as MTO, (D) 100,000lb, (E) 100,000lb, (F) as MTO, (FB-111A) as MTO, (FB-111H) 155,000lb, (EF) as MTO.

Engines

(F-111A to No 31 and YF) two Pratt & Whitney JTF10A-20/TF30-P-3, rated with full afterburner at 18,500lb each; (A, production) JTF10A-21/TF30-P-3, rated 18,500lb; (B) JTF10A/TF30-P-12, rated 20,250lb; (C) TF30-P-3; (D) JTF10A-36/TF30-P-9, rated 19,600lb; (E) TF30-P-3; (F) JTF10A-32C/TF30-P-100, rated 25,100lb; (EF) TF30-P-3; (FB-111A) JTF10A-27D/TF30-P-7, rated 20,350lb; (FB-111H) General Electric F101-GE-100, rated in 30,000lb class.

Performance

Maximum Mach number (sea level, clean, most versions) 1.1, (high altitude, clean, most versions) 2.2 (F, 2.5; H, 2+); normal maximum speed with maximum internal and external load (A,C,D,E) 490kt, 564mph; combat radius (A,C,D,E,F, Hi-Lo-Hi with 12,000lb bomb load on wings without air refuelling) 600-650 miles, (FB-111A with four tanks and internal SRAMs, without air refuelling) 1,800 miles; range (A,C,D,E,F, without air refuelling) about 3,000 miles, (FB-111A with air refuelling including 1,382-mile low-level high-speed dash) 6,100 miles, (FB-111H, same conditions) 'more than twice the FB-111A sea-level dash range with increased payload'; takeoff distance (A,C,D,E) 3,500-4,000ft, (F) 3,100ft, (FB-111A) 7,400ft, (FB-111H) 6,650ft.

Flight equipment

J-band multimode attack radar (A,C,E) General Electric APQ-113 (a few later had APQ-119), (B) Hughes AWG-9, (D) Autonetics APQ-130, (F,FB-111H) General Electric APQ-144, (FB-111A) General Electric APQ-114, (EF) none; TFR (most) Texas Instruments APQ-110, (D) Sperry APQ-128, (FB-111A and H) APQ-134; nav/attack subsystem, with inertial navigation system (all) Litton Industries, with AJN-16 INS; doppler radar (all except D) Singer/General Precision (FB-111A, APQ-185; FB-111H, APN-200), (D) Canadian Marconi; (FB-111A only) ASQ-119; lead-computing sight and optical display sight (tactical versions) General Electric; head-up display (D,F only) Norden Division of United Technologies; computer complex (A,C,E) Litton, (D,F, FB,EF) IBM; radar altimeters (Hi) Stewart-Warner, (Lo) Honeywell.

Note: Radar of the F-111F fitted with the Pave Tack AVQ-26 IR/laser pod under the weapon bay is modified to APQ-161. First Pave Tack Maverick launch took place at McClellan AFB in February 1978.

Mission equipment

Radar homing and warning (including SEESAM) (A,C) Loral ALR-31 (later ALR-39) and Bendix APS-107, (C,E) Loral ALR-41 and Dalmo Victor APS-109, (D,F,FB and EF) includes Dalmo Victor ALR-62; ECM noise jammer (tactical versions, usual fit) Sanders ALQ-109; ECM deception jammer (A,C,E) Sanders ALQ-94, (D,F,FB) Sanders ALQ-137, (FB-111H) advanced ALQ-137 plus new ALQ-131 for early-warning cover, (EF) Sanders ALQ-100 (mod) plus complete AIL ALQ-99A Tactical Jamming System; ECM receiver (most) Transco ALR-4, (B) Avco ALR-2; IR receiver (A) Texas Instruments, (others) Cincinnati Electronics; chaff dispenser (most) ALE-23 or (D,F) General Dynamics ALE-28 pneumatic; chaff/flare pod (when carried) Lundy ALE-44; reconnaissance equipment (RF) Fairchild and Hycon cameras, Westinghouse APD-8 SLAR (Side-Looking Airborne Radar) and Rockwell computer group; SRAM master computer (FB) Rockwell Autonetics.

Weapon loads

(F-111A,C,D,E and F)

In internal bomb bay: two nuclear weapons (various tactical), or two 2,000lb Mk 84 bombs, or M61A1 gun and ammunition, or ferry fuel tank. On left or right belly pylon (not both): one 600 USgal (500gal) drop tank.

On each inboard No 4 or 5 wing pylon (Sta.118 — inches from centreline): six Mk 82 GP bombs (nominal 500lb) or equivalent LDGP, HD or similar, or four M117R or 117D (750lb), or one Mk 84 (2,000lb) or one M118 (3,000lb) or one 500gal tank.

On each No 3 or 6 pylon (Sta.189): Same as Nos 4 and 5.

On each No 2 or 7 pylon (Sta.256): One 500gal tank, without other external stores (not normally fitted).

On each No 1 or 8 pylon (Sta.309): Not normally fitted.

The true weight of a typical Mk 82 is 580lb and an M117 about 825lb (larger bombs are heavier than the nominal figure in a similar proportion), so with pylons 3,4,5 and 6 fully loaded the external ordance mass is roughly 13,200-13,920lb.

Note: in mid-1977 development began intended to furnish the F-111F force with EO (electro-optical) sensors and guided ASM (Maverick) capability.

(FB-111A)

In weapon bay: two SRAM or B43, B61 or B77 nuclear bombs.

On pylons 3,4,5 and 6: one SRAM or 500gal tank.

(FB-111H)

In weapon bay: four SRAM, three B43, five B61 or four B77.

On fuselage pylons 1 to 6 (new): six SRAM, B43/61/77, ALCM, ALCM-B or TALCM.

On pylons 4 and 5: one SRAM or one B43/61/77.

On pylons 3 and 6: one SRAM, ALCM, ALCM-B, TALCM or B43/61/77.

On pylons 2 and 7: one ALCM, ALCM-B or TALCM.